THE INTERNATIONAL SCIENTIFIC SERIES.

VOLUME XX.

THE INTERNATIONAL SCIENTIFIC SERIES.

Works already Published.

I. FORMS OF WATER, IN CLOUDS, RAIN, RIVERS, ICE, AND GLACIERS. By Prof. JOHN TYNDALL, LL. D., F. R. S. 1 vol. Cloth. Price, $1.50.

II. PHYSICS AND POLITICS; OR, THOUGHTS ON THE APPLICATION OF THE PRINCIPLES OF "NATURAL SELECTION" AND "INHERITANCE" TO POLITICAL SOCIETY. By WALTER BAGEHOT, Esq., author of "The English Constitution." 1 vol. Cloth. Price, $1.50.

III. FOODS. By EDWARD SMITH, M. D., LL. B., F. R. S. 1 vol. Cloth. Price, $1.75.

IV MIND AND BODY: THE THEORIES OF THEIR RELATIONS. By ALEX. BAIN, LL. D., Professor of Logic in the University of Aberdeen. 1 vol., 12mo. Cloth. Price, $1.50.

V. THE STUDY OF SOCIOLOGY. By HERBERT SPENCER. Price, $1.50.

VI. THE NEW CHEMISTRY. By Prof. JOSIAH P. COOKE, Jr., of Harvard University. 1 vol., 12mo. Cloth. Price, $2.00.

VII. THE CONSERVATION OF ENERGY. By Prof. BALFOUR STEWART, LL. D., F. R. S. 1 vol., 12mo. Cloth. Price, $1.50.

VIII ANIMAL LOCOMOTION; OR, WALKING, SWIMMING, AND FLYING, WITH A DISSERTATION ON AËRONAUTICS. By J. BELL PETTIGREW, M. D., F. R. S. E., F. R. C. P. E. 1 vol., 12mo. Fully illustrated. Price, $1.75.

IX. RESPONSIBILITY IN MENTAL DISEASE. By HENRY MAUDSLEY, M. D. 1 vol., 12mo. Cloth. Price, $1.50.

X. THE SCIENCE OF LAW. By Prof. SHELDON AMOS. 1 vol., 12mo. Cloth. Price, $1.75.

XI. ANIMAL MECHANISM. A TREATISE ON TERRESTRIAL AND AËRIAL LOCOMOTION. By E. J. MAREY. With 117 Illustrations. Price, $1.75.

XII. THE HISTORY OF THE CONFLICT BETWEEN RELIGION AND SCIENCE. By JOHN WM. DRAPER, M. D., LL. D., author of "The Intellectual Development of Europe." Price, $1.75.

XIII. THE DOCTRINE OF DESCENT, AND DARWINISM. By Prof. OSCAR SCHMIDT, Strasburg University. Price, $1.50.

XIV. THE CHEMISTRY OF LIGHT AND PHOTOGRAPHY. IN ITS APPLICATION TO ART, SCIENCE, AND INDUSTRY. By Dr. HERMANN VOGEL. 100 Illustrations. Price, $2.00.

XV. FUNGI; THEIR NATURE, INFLUENCE, AND USES. By M. C. COOKE, M. A., LL. D. Edited by Rev. M. J. BERKELEY, M. A., F. L. S. With 109 Illustrations. Price, $1.50.

XVI. THE LIFE AND GROWTH OF LANGUAGE. By Prof. W. D. WHITNEY, of Yale College. Price, $1.50.

XVII. MONEY AND THE MECHANISM OF EXCHANGE. By W. STANLEY JEVONS, M. A., F. R. S., Professor of Logic and Political Economy in the Owens College, Manchester. Price, $1.75.

XVIII. THE NATURE OF LIGHT, WITH A GENERAL ACCOUNT OF PHYSICAL OPTICS. By Dr. EUGENE LOMMEL, Professor of Physics in the University of Erlangen. With 188 Illustrations and a Plate of Spectra in Chromolithography. Price, $2.00.

XIX. ANIMAL PARASITES AND MESSMATES. By Monsieur VAN BENEDEN, Professor of the University of Louvain, Correspondent of the Institute of France. With 83 Illustrations. Price, $1.50.

THE INTERNATIONAL SCIENTIFIC SERIES

ON

FERMENTATION

BY

P. SCHÜTZENBERGER

DIRECTOR AT THE CHEMICAL LABORATORY AT THE SORBONNE

WITH TWENTY-EIGHT ILLUSTRATIONS

NEW YORK:

D. APPLETON AND COMPANY,

549 AND 551 BROADWAY.

1876.

CONTENTS.

BOOK II.

ALBUMINOID SUBSTANCES—SOLUBLE OR INDIRECT FERMENTS—ORIGIN OF FERMENTS.

LIST OF ILLUSTRATIONS.

NOTE BY TRANSLATOR.

The French words "invertir," "inverti," "intervertir," "inversive," &c., p. 28, &c., may be rendered in English by the recognized terms, "inverted," "inversive," &c. Yet since these terms are not understood by all, and present some difficulty, it has been thought better to adopt in the text the less technical and more suggestive rendering "altered," "alterative," &c.; by altered sugar being meant cane-sugar which has taken up a molecule of water and split up into a mixture of glucose and levulose.

ON FERMENTATION.

INTRODUCTION.

FERMENTATION is only a particular instance, selected from among the chemical phenomena of which living organisms are the field; it, like all biological reactions, comes before us as a manifestation of the special force residing in these organisms, or rather in their cellular elements.

If we leave the nature of the fermenting body, and the products derived from it, in the background, there is nothing to distinguish fermentation from the other chemical transformations which take place in the animal or vegetable economy.

The reason why the production of alcohol and carbon dioxide at the expense of sugar, the conversion of glucose into lactic and butyric acids, and other phenomena of the same order have been classed by themselves, is that the real cause of these curious transformations was long misunderstood. It had not been observed that they had as their origin the presence of living organisms, or, at least, principles which are directly derived therefrom.

There is, therefore, no longer any necessity, in the present state of science, for grouping together under a special name these various reactions ; it is more convenient, on the contrary, to class them among the general mass of chemical phenomena of the living organism.

We must, consequently, do one of two things ; either cease to use the term fermentation, as a general expression applying to a certain order of phenomena, or we must designate by it all those changes which, by the special conditions under which they are produced, are evidently due to the intervention of a force differing from those which we handle in our laboratories.

It is true that the organisms which give rise to what have been hitherto called fermentation are simple elementary organisms reduced to a single cell ; but a plant or an animal of a high order is only the union, under special laws, of different kinds of cells, each of which acts in a certain determinable manner. When, as M. Pasteur has remarked, we sow at the same time, in the same saccharine medium, alcoholic, lactic, and butyric ferment, we see three distinct reactions take place, one of which splits up the sugar into alcohol and carbon dioxide, the second converts it into lactic acid, and the third into butyric acid.

The more simple an organism is, the fewer special kinds of cells it contains, the simpler are the chemical reactions which take place in it, and the more easily are they separated from each other, and isolated by experiment. On the contrary, in proportion as the histological constitution is varied and heterogeneous, we see a greater number of distinct compounds, as the products of the many chemical changes which take place in the different tissues.

As a consequence of what we have just said, our plan would be considerably enlarged, and the history of fermentation would become that of the chemical phenomena of life.

We will not, however, give such a wide scope to this work, but will confine ourselves to the examination of the phenomena which have been hitherto designated by the name of fermentation. Under these restrictions, the history of fermentation may be considered as an introduction to biological chemistry.

In fact, it is easily seen, from the preceding considerations, that the thorough study of ferments, properly so called, or rather of elementary organisms, and of their mode of existence, ought to precede that of the more complete beings. We more easily understand the properties of granite, and the influence exerted upon it by water and atmospheric agents, when we have learned that it is formed of crystals of quartz, felspar, and mica in juxtaposition, and have studied the chemical characters of each of these compounds. In the same manner, the study of the chemical manifestations of the vital force in cellular organisms is destined to throw a bright light on the more complex functions of the higher plants and animals. This has been recognized by M. Pasteur, and by all those who have subsequently entered on the physiological study of fermentation, and of the development of cellular organisms.

A living cell of beer-yeast possesses the property of resolving into alcohol, glycerin, carbon dioxide, and succinic acid the altered sugar which penetrates by endosmose through its membranous envelope.

If we substitute for the cell of beer-yeast a cell of lactic ferment, we still see the sugar disappear, but the products into which the ponderable elements of the glucose are resolved are different; instead of alcohol and carbon dioxide, we have lactic acid. The *modus faciendi* of the vital force of this cell is evidently not the same as that of the former one; but we cannot, therefore, affirm that there are as many vital chemical forces as there are reactions.

When a pencil of solar light passes through a prism, the constituent parts of this pencil are isolated on account of their unequal refrangibility. The least refrangible rays are revealed to us by the effects of heat (the dilatation and change of condition of bodies); next come the luminous rays, which excite on the retina the impressions of colour forming the spectrum; and then, beyond the violet, is a series of invisible rays, which are revealed only by their decomposing action on certain combinations (salts of silver, &c.). But we know now that all these calorific, luminous, and chemical rays, some of which give heat without light, and others light without heat, or excite chemical reactions, differ only in the rapidity of the vibratory movements of the ether, and are essentially distinguishable from each other only by their wave-length. It is possible that an analogous bond may unite the vital chemical forces of the different elementary organisms. Sand, sprinkled uniformly on the surface of a vibrating plate, collects in nodal lines of different forms, according to the sharpness of the note which we draw from this plate by means of a bow; in the same manner, chemical compounds may perhaps be resolved into more simple combinations, varying in kind according to the vibratory rhythm which starts them.

The transformation of sugar into alcohol and carbon dioxide, and the conversion of the same body into lactic acid are chemical phenomena which we cannot yet reproduce by the intervention of heat alone, nor by the additional agency of light or of electricity. The force capable of attacking, in a certain determinate direction, the complex edifice which we call sugar, an edifice composed of atoms of carbon, hydrogen, and oxygen, grouped according to a determinate law—this force, which is manifested only in the living cell of the ferment, is a force as material as all those which we are accustomed to utilize.

Its principal peculiarity is, that it is only found in the living organisms, to which it gives their peculiar character. We ought not to allow ourselves to be stopped by this rampart, over which no one has hitherto been able to pass; we ought not to say to the chemist, "You shall go no farther, for beyond this is the domain of life, where you have no control."

The history of science shows us the weakness of these so-called impassable barriers.

Gerhardt, when he published his excellent treatise on organic chemistry, thought himself justified in saying, "It is vital force alone which acts synthetically and reconstructs the edifice demolished by chemical forces."

M. Berthelot, some years afterwards, in a brilliant series of discoveries, made the first successful attempt to perform organic syntheses, and determined the principal conditions under which they can be effected.

In a remarkable lecture on molecular dissymmetry (Leçons de la Société Chimique de Paris, 1860), M. Pasteur had established an important distinction between artificial organic products and the compounds formed under the influence of living organisms.

"The artificial products of the laboratory have coincident images (sont à image superposable). On the contrary, most of the natural organic products—I might say *all*, if I had only to allude to those which play an important part in the phenomena of vegetable and animal life—all the products essential to life are unsymmetrical, and unsymmetrical in such a way that their images cannot be made to coincide with them." And afterwards he says, "We have not yet been able to realize the production of an unsymmetrical body, by the aid of compounds which are not so themselves."

Nearly at the same time that these words were uttered before the Chemical Society of Paris, two English chemists, Perkin and Duppa, succeeded in transforming succinic acid into tartaric acid. M. Pasteur himself acknowledged that the artificial product of Perkin was a mixture of paratartaric acid and of inactive tartaric acid. But paratartaric acid easily splits up, as Pasteur's elegant experiments have shown, into dextro-tartaric and lævo-tartaric acid, and M. Jungfleisch has shown that inactive tartaric acid, heated with water at 175°, is partially converted into paratartaric acid.

The succinic acid employed by the English chemists was formed by the oxidation of yellow amber. This was not a synthetical product; it might be thought that, though it was inactive, it resulted, like racemic acid, from the union of two active but opposed molecules. Jungfleisch has removed this last doubt. He prepared synthetic succinic acid by a well-known method, by means of cyanide of ethylene and potassium. This acid furnished paratartaric acid, like that produced from amber.

Thus fell the barrier placed by M. Pasteur between natural and artificial products. This example shows us how cautious we ought to be in making distinctions which we seem justified in establishing between the chemical reactions of the living organism and those of the laboratory. Because a chemical phenomenon may hitherto have been produced only under the influence of life, it does not follow that it will never be effected otherwise.

No one can any longer admit that vital force has power over matter, to change, counterbalance, or annul the natural play of chemical affinities. That which we have agreed to call chemical affinity is not an absolute force; this affinity is modified in numberless ways, according as the circumstances by which bodies are surrounded, vary. Thus, the apparent differences between the reactions of the laboratory and those of the organism ought to be sought for, more particularly among the *special conditions*, which the latter alone has been able hitherto to bring together.

In other words, there is really no chemical vital force. If living cells produce reactions which seem peculiar to themselves, it is because they realize conditions of molecular mechanism which we have not hitherto succeeded in tracing, but which we shall, *without doubt*, be able to discover at some future time. Science can gain nothing by being limited in the possibility of the aims which she proposes to herself, or the end which she seeks.

If, in this work, we still employ the expression, "the vital chemical force of an elementary organism," it will be clearly understood that we intend these words to signify the realization of the conditions of molecular mechanism necessary in order to set up a certain reaction.

We will not delay any longer over these general considerations, which are, after all, nothing but hypotheses, naturally suggesting themselves to the mind of him who seeks to explain the causes which produce certain observed effects, but on which it is not necessary to dwell at the present stage of our inquiry; we will therefore proceed at once to the examination of facts.

The study of fermentation may be divided into two parts, according to the nature of the ferment. The first will comprise the fermentation due to the intervention of an organized ferment, having a determinate form; the second will be reserved for fermentation produced by soluble products, elaborated by living organisms.

BOOK I.

FERMENTATION DUE TO CELLULAR ORGANISMS, OR DIRECT FERMENTATION.

CHAPTER I.

HISTORICAL.

THE word fermentation is derived from *fervere*, to boil; it evidently owes its origin to the reaction presented by saccharine liquids, when they are left to themselves or placed in contact with ferments. We observe, in fact, in this case, a more or less abundant disengagement of gas, which causes the liquid to effervesce or boil. The sugar disappears, and the product becomes spirituous. The expression, fermentation, was subsequently applied to other phenomena, in which an organic body, when dissolved, is modified, changed, and transformed, under the influence of a cause which remained for a long time unknown and badly defined. Thus the acidification of wine was called fermentation, although in this case there was no effervescence. The analogy of the determining cause was considered, rather than the appearance of the phenomenon.

Alcoholic fermentation was the first known, and was

also more studied than the other reactions of this class. Osiris among the Egyptians, Bacchus among the Greeks, Noah, according to the Israelitish tradition, taught men the art of cultivating the vine, and making wine. Moses, in his writings, draws a distinction between unleavened and leavened bread, and relates that the Israelites were in such haste, during their flight from Egypt, that they had no time to put leaven into their dough. The ancients used as leaven for their bread either dough that had been kept till it was sour, or beer-yeast.

"Galliæ et Hispaniæ frumento in potum resoluto, spuma ita concretâ pro fermento utuntur, quâ de causa levior illis, quam ceteris, panis est," says Pliny, who also adds, that in the fermentation of bread acidity is the most active principle. From the earliest times certain fermented liquids were known, both in Egypt and Germany, prepared from natural saccharine juices which had been allowed to ferment; such as beer, hydromel, palm-wine, and cider.

We find, in short, from all ancient documents, that alcoholic fermentation was empirically known in its principal effects, and utilized at a period far earlier than that which has left written traces of its history.

Among the writings of the alchemists from the thirteenth to the fifteenth century, we very frequently find the expressions "fermentation and ferments" (fermentatos et fermentum), without our being able to ascertain clearly what precise ideas they attached to them. They knew no distinction between mineral and organic substances; the phenomena connected with the changes in organic products were assimilated and confounded with the transformations of mineral compounds, and with the solution of salts and metals. The term "ferment" was often applied even to the philosopher's stone.

"Apud philosophos fermentum dupliciteo videtur dici; uno modo ipse lapis philosophorum et suis elementis compositus, et completus in comparatione ad metalla; alio modo illud, quod est perficiens lapidem et ipsum complens.

"De primo modo dicimus, quod sicut fermentum pastæ vincit pastam, et ad se convertit semper, sic et lapis convertit ad se metalla reliqua. Et sicut una pars fermenti pastæ habet convertere partes pastæ et non converti, sic et hic lapis habet convertere plurimas partes metallorum ad se, et non converti."—PETRUS BONUS *of Ferrara*, 1330-1340.

We see that the writer is especially struck with this fact, that a very small quantity of leaven transforms into fresh leaven an almost indefinite quantity of paste. This property of transmitting a force to a large mass without being itself weakened by the process, was precisely that which ought to characterize the philosopher's stone which was so much sought after.

Basil Valentine, in his "Triumphal Car of Antimony," admits that yeast, employed in the preparation of beer, communicates to the liquor an internal inflammation, and determines thereby a purification, and a separation of the clear parts from those which are troubled.

Alcohol, the presence of which in the fermented liquid was known to him, was considered by him to exist previously in the decoction of germinated barley; but that it did not become active and susceptible of being separated by distillation until it had been cleared from the impurities which accompany it, and mask its special properties.

Libavius (Alchymià, 1595) believed that, "Fermentatio est rei in substantia, per admistionem fermente quod virtute per spiritum distributo totam penetrat massam et in suam naturam immutat, exaltatis." The ferment must be of a similar nature to the matter which enters into fermentation, and the latter must be either liquid, or in a state of minute division; the principal agent resides in the heat of the ferment.

Like the chemists of a later age, Libavius compares fermentation to putrefaction, and considers them as different effects of the same cause. He protests, on the contrary, against the confused ideas which had been entertained concerning digestion and fermentation. Digestion is, according to him, "motus ad mistionem, non ad perfectionem," as fermentation is.

The iatro-chemical school attributed to fermentation a preponderating power, and even confounded under this term a great number of chemical reactions.

Thus Van Helmont expresses himself as follows in his "Ortus Medicinæ" (648): "Docebo omnem transmutationem formalem præsupponere fermentum corruptivum."

The formation of intestinal gasses, the production of blood and animal fluids, spontaneous generation, the effervescence of chalk under the influence of acids, are phenomena with which fermentation has to do.

We may, however, say in passing that Van Helmont had the merit of clearly distinguishing the production of a special gas (gas vinorum) during alcoholic fermentation. He says that this gas vinorum is different from spirit of wine, as he was able to prove by experiments. It is impossible to discover from his writings whether or no he recognized the identity of the gas vinorum and the gas carbonum produced during the combustion of charcoal.

In 1664, Wren pointed out that the gas produced by alcoholic fermentation can be absorbed by water, like that which is disengaged by the action of an acid on salt of tartar.

Silvius de la Boë (1659) no longer regarded the effervescence of the alkaline carbonates under the influence of acids as a phenomenon of the same class as fermentation.

He supposed that in the former case there was combination, in the latter decomposition.

Lemery (Cours de Chimie, 1675) is not so explicit when he says: "The fermentation which occurs in paste, wort, and all other similar things, is different from that of which we have just spoken (effervescence), since it is slower; it is excited by the natural acid salt of these substances, which, becoming disengaged, and having its energy increased by its motion, rarefies and raises the gross and oily part which opposes its passage, and thus we see the matter rise.

"The reason why the acid does not cause sulphurous substances to ferment with as much noise and readiness as alkalis, is that oils are composed of pliant parts which yield to the points of the acid, as a piece of wool or cotton would yield to needles pressed into it. Thus it seems to me that we must admit of two sorts of fermentations; one of acids with alkalis, which would be called effervescence; and the other would be, when the acid rarefies by degrees a solid matter like paste, or clear and sulphurous like wort, cider, or other juices of plants; we should call the latter sort fermentation."

Lemery says again, when speaking of alcoholic fermentation: "In order to understand this effect, we must know that wort contains much essential salt; this salt, being volatile, makes an effort, during fermentation, to detach itself from the oily particles by which it is, as it were, bound; it penetrates them, divides and separates them, until by its subtile and piercing points, it has rarefied them into spirit; this effort causes the ebullition which takes place in wine, and, at the same time, its purification; for it separates and removes the grosser parts in the form of froth, a portion of which attaches itself to the sides of the vessel and grows hard; the other falls to the bottom, and is called tartar and lees. The inflammable spirit of wine is therefore nothing but an oil exalted, that is purified, by salt."

We find in the researches and writings of Becker (1682), a very marked progress in the study of the products of fermentation. He was the first to bring forward the important fact that saccharine liquids alone are capable of entering into spirituous fermentation. He considers that alcohol does not pre-exist in the wort, but is formed during the process of fermentation; the intervention of air is necessary to set this action going, which he considers analogous to combustion. Becker brings together, under the name of fermentation, the production of gas in the stomach of sick animals (insumefactio), spirituous fermentation (proprie fermentatio), and acetification (acetificatio seu acescentia).

We owe to Willis (1659), and to Stahl, the celebrated originator of the theory of phlogiston (1697), the first philosophical conception of the peculiar nature of fermentation, or rather of *fermentations*. According to their views a ferment is a body endued with a motion peculiar to itself, and it transmits this motion to the fermentable matter. Thus Willis says in his dissertation "De Fermentatione":—

"Fermentatio est motus intestinus cujusvis corporis, cum tendentia ad perfectionem ejusdem corporis vel propter mutationem in aliud. Plures sunt modi quibus fermentatio promooctur. Primus et principius erit fermenti cujusdam corpori fermentando adjectio; cujus particulæ cum prius sint in vigore et motu positæ, alias in massa fermentanda otiosas et torpidas exsuscitant, et in motum vindicant."

Stahl considered alcoholic fermentation as a phenomenon of the same class as putrefaction, and as only a particular case of it. As there was at that time no definite idea of the elementary composition of fermentable substances, and of the products of their fermentation, there evidently could not be established any correct relation between these bodies, and any hypothesis could be safely brought forward. Thus Stahl considers that fermentable matter (sugar, flour, milk) is composed of particles formed by the unstable union of salt, oil, and earth; under the influence of the internal motion excited by the ferment, the heterogeneous particles are separated from each other, and then recombined so as to form more stable compounds including the same principles, but in other proportions.

From Stahl to Lavoisier we find no names of great

note, and no interesting discoveries with respect to fermentation.

When chemistry underwent its great transformation at the end of the last century, under the powerful influence of the genius of Lavoisier, fermentation necessarily attracted anew the attention of experimentalists. Lavoisier himself studied it (Elemens de Chimie, vol. i. p. 139, second edition), and as was the case with all subjects which he handled, he threw a ray of light upon the darkness. Proceeding in his usual manner, balance in hand, by weight and measure, and applying to the solution of the problem the new methods of organic analysis which he had invented, he endeavoured to ascertain the bond or relation which exists between the fermented matter, the sugar, and the products of fermentation, alcohol and carbon dioxide.

From this moment we quit the domain of the history of the science, and enter into that of the real and well-observed facts which will be treated of in the following chapters.

We may say, in recapitulation, that, before the labours of Lavoisier and his followers, the fermentable products and the principal terms of their transformations (carbon dioxide gas, alcohol, acetic acid, &c.) were known *qualitatively*. The distinction between the acid, or acetic fermentation, and the alcoholic fermentation was known; there was an idea of the analogy which exists between putrefaction and alcoholic fermentation; and an explanation of the manner in which a ferment acts had been sought.

The latter was known only as a kind of foam, deposit, or paste, in which resided an occult and special force, capable of determining chemical phenomena. We may add that these phenomena were considered as distinct, both in their action and exciting cause, from the ordinary reactions of chemistry. This was, as one may see, but a slight result of the many volumes that had been written on this subject.

Spirituous or alcoholic fermentation being, in every respect, the part of this subject which has been the most thoroughly studied, we will commence our monograph by its examination.

CHAPTER II.

ALCOHOLIC OR SPIRITUOUS FERMENTATION.

PASTEUR, in his excellent memoir (Ann. de Chimie et Physique, 3rd series, vol. lviii. p. 323), calls by the name of *alchoholic fermentation* that which sugar undergoes under the influence of the ferment which bears the name of yeast or barm.

We can only adopt this definition as applying, without any possibility of uncertainty, to a phenomenon very limited in its cause and its effects; but we shall have to inquire, in a later portion of this work, whether alcohol cannot be produced at the expense of sugar under other influences than those of the product known as beer-yeast.

As we have before said, the splitting up of a molecule of sugar into many more simple products, among which we find alcohol and carbon dioxide, is the consequence of a special mechanical action, exercised on the ultimate particles of the compound matter. Whatever may be the source, whether living organism or dead matter, which realizes the conditions necessary for this rupture of equilibrium, the phenomenon will be essentially the same. In a general and philosophical point of view, there is no more reason why we should separate alcoholic fermentation excited by yeast from that which is due to any other agent, than why we should distinguish cane sugar from that produced from beet-root.

While we restrict, with Pasteur, the expression "alcoholic fermentation," and do not include in it all the phenomena of decomposition, in which alcohol is produced, we have to consider the body which ferments, the sugar, or rather the sugars, the products of fermentation, among which alcohol takes its place in the first rank, and then the determining cause of the fermentation of sugar, beer-yeast.

PRODUCTS OF THE REACTION.

Let us first consider alcoholic fermentation as an ordinary chemical reaction ; in other words, let us study it by means of the body which is decomposed, and the products which are derived from it; we will then consider the cause of the decomposition, and the properties of this ferment, as well as those of analogous products.

We said before that Becker was the first to recognize the necessity of the presence of sugar in wines which undergo spirituous fermentation, but that to Lavoisier belongs the honour of having studied and demonstrated the relations of their composition which connect sugar with its derivatives.

Setting out with this principle, that nothing is created either in the operations of art, or in those of nature; that in every operation there is an equal quantity of matter both before and after the operation ; that the quality and quantity of the elements are the same, and that there are only changes and modifications, this illustrious chemist established by analysis the centesimal proportions of carbon, hydrogen, and oxygen contained in sugar, operating in the same manner on the alcohol, the carbon dioxide, and acetic acid recognized by him as the products of the decomposition of sugar; then, estimating by analysis the respective quantities of these three bodies which are formed at the expense of a known weight of sugar, he ascertained the result of the reaction, and arrived at the following conclusions:—

"The effects of vinous fermentation are reduced to separating into two portions the sugar, which is an oxide, and oxidizing one at the expense of the other, so as to form from it carbon dioxide, in reducing the other in favour of the former, to form from it a combustible substance, alcohol; so that if it were possible to recombine these two substances, alcohol and carbon dioxide, we should reform the sugar." He had really made a great advance on the conceptions of Stahl, founded on a mixture of salt, oil, and earth.

The researches of Lavoisier may be summed up by the following equation:—

95·9 parts of crystallized cane sugar contain 26·8 of carbon, 7·7 of hydrogen, and 61·4 of oxygen.

These are decomposed, and form 57·7 parts of alcohol, containing 16·7 carbon, 9·6 hydrogen, and 31·4 oxygen + 35·3 parts of carbon dioxide, containing 9·9 carbon, and 25·4 oxygen + 2·5 parts of acetic acid, containing 0·6 carbon, 0·2 hydrogen, and 1·7 oxygen.

Thus we find :—

Carbon of the sugar	26·8
Hydrogen „ „	7·7
Oxygen „ „	61·4
Sum of carbon of the three products	27·2
„ „ hydrogen „ „	9·8
„ „ oxygen „ „	58·5

Taking into consideration the imperfections of his method of analysis, Lavoisier thought the agreement between the two sides of this equation sufficient to confirm the general principle announced above.

If we compare with these numbers those furnished by the wonderfully accurate methods employed by modern chemists, we shall see that, in reality, 95·9 parts of cane sugar contain :—

44·4 carbon, 6·1 hydrogen, and 49·4 oxygen;

and give,

51·6 of alcohol, containing—

26·9 carbon, 6·7 hydrogen and 18·0 oxygen +

49·4 parts of carbon dioxide, containing—

13·5 carbon, and 36·9 oxygen.

It was therefore only by compensation of considerable errors that Lavoisier was led to an approximate solution.

Towards 1815, the analyses so carefully made by Gay-Lussac and Thénard, and by De Saussure, settled in a determinate manner the composition of sugar and of alcohol. These results, far from invalidating the conclusions of Lavoisier, gave them solid support. Thus Gay-Lussac (Ann. de Chimie, vol. 95, p. 318) wrote: "If it be supposed, now, that the products furnished by the ferment can be neglected, as far as relates to the alcohol and carbonic acid which are the only sensible results of fermentation, it will be found that, given 100 parts of sugar, 51·34 of them will be converted during fermentation into alcohol, and 48·66 into carbonic acid."

These results, expressed in a chemical equation,* give to cane sugar the formula $C_{12} H_{24} O_{12}$,† and we shall have $C_{12} H_{24} O_{12} = 4 C_2 H_6 O + 4 CO_2$.‡ The analyses of cane sugar made by Gay-Lussac and Thénard themselves agree, as well as those since made by a great number of chemists, with the formula $C_{12} H_{22} O_{11}$.§

In order to arrive at the error which we have just pointed out, and which Messrs. Dumas and Boullay showed in 1828, Gay-Lussac supposed that his analyses of cane sugar were imperfect, and he modified them recklessly, in the proportion of 2 or 3 per cent., in order to establish an agreement between the two sides of his equation.

"The theory of fermentation arrived at by Gay-Lussac is still imperfect," said Messrs. Dumas and Boullay, "but it is no longer so when we substitute ether for alcohol in the theoretical composition of sugar. The most complete agreement is then established between theory and experiment." The conclusion which these two chemists deduced from this obser-

* These formulæ are given by M. Schützenberger as original formulæ. It may be worth while, from an historical point of view, to preserve them as written by their enunciators, in the "old notation," thus :—

† $C_{12} H_{12} O_{12}$. ‡ $C_{12} H_{12} O_{12} = 2 C_4 H_6 O_2 + 4 CO_2$.
§ $C_{12} H_{11} O_{11}$.

vation, was that cane sugar cannot ferment without assimilating the elements of a molecule of water. In other words, Gay-Lussac's equation, as a numerical expression, is correct, but that it would be better to write the first member of it under the form—

$$\underset{\text{cane sugar}}{C_{12} H_{22} O_{11}} + \underset{\text{water}}{H_2 O} = \underset{\text{alcohol}}{4 C_2 H_6 O} + \underset{\text{carbon dioxide.}}{4 CO_2}.*$$

A little later (1832), Dubrunfant observed that before fermentation commenced, the cane sugar is transformed into uncrystallizable sugar.

M. Berthelot proved afterwards that the taking up of water by the cane sugar which precedes alcoholic fermentation is due to the presence of a soluble ferment in the yeast; we shall return again to this important point. Finally, in 1833, Biot discovered the change of sugar under the influence of acids.

Gay-Lussac's equation, modified by Dumas and Boullay, was generally admitted for more than twenty years, as the mathematical expression of the decomposition of sugar by yeast.

Meanwhile, in 1856, Dubrunfant, by making a quantitative analysis of the carbon dioxide disengaged by fermentation, observed that it was not possible to make experimentally the equation of fermentable sugars with alcohol and carbon dioxide only. (Comp. Rend. de l'Acad. des Sciences, vol. 42, p. 945.)

The latest important work on the qualitative and quantitative analyses of the products of the alcoholic fermentation of sugars is due to M. Pasteur. (Ann.

* $\underset{\text{cane sugar}}{C_{12} H_{11} O_{11}} + \underset{\text{water}}{HO} = \underset{\text{alcohol}}{2 C_4 H_6 O_2} + \underset{\text{carbon dioxide.}}{4 CO_2}.$

Chimie et Physique, 3rd series, vol. 58, p. 330.) By a series of very interesting researches, and by irrefutable experiments, this illustrious chemist proves: 1st. That in every alcoholic fermentation, besides the principal products, alcohol and carbon dioxide, glycerin and succinic acid are formed; 2nd. That the glycerin and succinic acid are produced at the expense of the elements of the sugar, and that the ferment takes no part in it; 3rd. That, besides this, the sugar yields a certain portion of its substance to the new ferment which is developed; we shall return to the last point when we more especially consider the ferment; 4th. That the lactic acid, the production of which, in variable quantities, has been observed in alcoholic fermentation, is the result of a special fermentation, differing from alcoholic fermentation, and proceeding simultaneously with it.

Let us say, in conclusion, in order to give to every one his due, that the presence of succinic acid in fermented liquids had been observed, before M. Pasteur, by Dr. Schmidt of Dorpat (Handwörterbuch der Chimie, Von Liebig, Poggendorff, 1st edit., vol. 3, p. 224, 1848), and also by Schunck in the fermentation of sugar by means of erythrozyme, the ferment derived from madder.

These facts had passed unperceived, and had been forgotten, at the time when Pasteur returned to the study of this subject.

Without entering into the details of the experiments on which Pasteur's conclusions rest, and which will be found in his memoir (loco citato), we will simply give the results of his quantitative researches.

100 parts of cane sugar $C_{12} H_{22} O_{11}$,* corresponding with 105·26 grape sugar 2 $C_6 H_{12} O_6$,† give nearly,—

Alcohol	51·11	
Carbon dioxide	48·89	according to Gay-Lussac's equation
	0·53	excess over " "
Succinic acid	0·67	
Glycerin	3·16	
Matter united with ferments	1·00	
	100·00	

Thus, out of 100 parts of cane sugar, about 95 parts are decomposed, according to Gay-Lussac's equation; 4 parts disappear and form succinic acid, glycerin, and carbon dioxide, and 1 part is added to the newly-formed ferment.

Pasteur endeavours to represent by an equation the decomposition of the 4 parts of sugar, which yield succinic acid and glycerin. This expression is very complex :—

$$49\,(C_{12} H_{22} O_{11} + H_2 O)$$

$$\text{or } 49\,(C_{12} H_{24} O_{12}) + 60\,H_2 O = \underset{\text{succinic acid}}{24\,(C_4 H_6 O_4)} + \underset{\text{glycerin}}{144\,(C_3 H_8 O_3)} + \underset{\text{carbon dioxide}}{60\,CO_2}.‡$$

This equation can only be considered, as Pasteur himself says, as a very approximate expression of the numerical results of the analysis, and not as a mathematical expression of the reaction.

* $C_{12} H_{11} O_{11}$. † $C_{12} H_{12} O_{12}$.

‡ 49 $(C_{12} H_{11} O_{11} + HO)$, or 49 $(C_{12} H_{12} O_{12})$ + 60 HO = 12 $(C_8 H_6 O_8)$ + 72 $(C_6 H_8 O_6)$ + 60 CO_2.

M. Monoyer (Thèse de la Faculté de Médicine de Strasbourg) proposes a much more simple equation to represent Pasteur's analysis :—

$$4\ (C_{12}\ H_{22}\ O_{11} + H_2\ O)\ *$$

or $4\ (C_{12}\ H_{24}\ O_{12}) + 6\ H_2\ O = 2\ (C_4\ H_6\ O_4) + 12\ (C_3\ H_8\ O_3) + 4\ CO_2 + O_2$.

He supposes, at the same time, that the excess of oxygen serves for the respiration of the globules of the ferment, a very plausible interpretation, as we shall presently see.

In order to understand the chemical possibility of the production of glycerin and succinic acid at the expense of sugar, it is sufficient to remark that by adding together the formulæ of glycerin and succinic acid, atom by atom, we arrive at a sum in which hydrogen and oxygen are in the proportions to form water :—

$$C_4\ H_6\ O_4 + C_3\ H_8\ O_3 = C_7\ H_{14}\ O_7.$$

On the other side we have,

$$C_7\ H_{14}\ O_7 + H_2\ O = 2\ C_3\ H_8\ O_3 + C\ O_2.$$

These two equations explain naturally enough the formation of glycerin and succinic acid, at the expense of glucose.

Even according to the researches of Pasteur, the proportions of glycerin and succinic acid, in relation to the alcohol, furnished by the same weight of sugar, are not absolutely constant. More glycerin and succinic acid, and less alcohol are formed, according as the fermenta-

* $4\ (C_{12}\ H_{11}\ O_{11} + HO)$, or $4\ (C_{12}\ H_{12}\ O_{12}) + 6\ HO = C_8\ H_6\ O_8 + 6\ C_6\ H_8\ O_6 + 2\ C_2\ O_4 + O_2$.

tion is slower, or is made with more exhausted and less pure yeast, supplied with but few alimentary principles, and those unsuited for the multiplication of its globules.

Fermentation effected by sowing the ferment, in the presence of more than a sufficient quantity of albuminoid and mineral matter suited to the nature of the globules, furnishes less glycerin and succinic acid, and more alcohol.

A feeble acidity of the liquor seems also to diminish the proportion of the two secondary products; the contrary occurs if the medium be neutral. Yet Pasteur himself says, that we usually find in wines a very large proportion of glycerin and succinic acid, although the fermentation of the must of the grape takes place in an acid medium, in presence of a sufficient quantity of albuminoid and mineral matters.

Whatever these variations may be, it is not less true that glycerin and succinic acid were never deficient in more than a hundred analyses of fermentation made by Pasteur.

The following is the method followed by M. Pasteur, to detect and measure quantitatively the glycerin and succinic acid contained in a fermented liquid. The liquid, when the fermentation is over, and all the sugar has disappeared (which requires from fifteen to twenty days under good conditions), is passed through a filter, accurately weighed against another made of the same paper. After having been dried at 100° C. (212° F.), the dried deposit of the ferment which is collected at the bottom of the vessel is accurately weighed. The filtered liquid is subjected to a very slow evaporation (at the rate of from 12 to 20 hours for each half-litre).

When it is reduced to 10 or 20 cubic centimetres ('61 or 1'22 cub. in.), the evaporation is finished in a dry vacuum. The sirupy residuum in the capsule is treated several times with a mixture of alcohol and ether, formed of one part of alcohol at 30° or 32°, and 1½ parts of rectified ether. After six or seven washings, there remains no more succinic acid or glycerin. The etherized alcoholic liquid is distilled in a retort, then evaporated in a water-bath in a capsule, and afterwards in a dry vacuum. Pure lime water is added to the remainder, till it is neutralized; it is then evaporated afresh, and the dried mass is again treated with the mixture of alcohol and ether, which only dissolves out the glycerin, leaving the calcium succinate in the form of a crystallized powder, stained with a small quantity of extractive matter, and with an uncrystallizable salt of lime. The calcium succinate is easily purified by treating it with alcohol at 80 per cent., which only dissolves out the foreign matters. The etherized alcoholic solution of glycerin is evaporated, and weighed after dessication in a dry vacuum.

Among the products met with normally and constantly in all alcoholic fermentations of sugar, we ought to mention acetic acid. The formation of this body noticed by Béchamp (Comp. Rend. de l'Acad., 1863), was at first attributed by Pasteur to a concomitant or subsequent acetic fermentation, and to the presence of *Mycoderma aceti;* but since the very precise and conclusive researches of Duclaux (Thèses Présentées à la Faculté des Sciences, 1865), it has been established: 1st. That the acetic acid is never deficient, even in fermentations conducted in the most careful manner, in order to preserve them from contact with air; 2ndly. That the proportion of this acid is remarkably constant, especially if we are careful to stop the fermentation as soon as all the sugar is transformed; besides this, it does not exceed ·05 per cent. of the weight of the sugar. This proportion of acetic acid is considerably augmented if the fermentation be continued beyond the limits just indicated. As we shall presently see that the ferment, when left to itself, without sugar and without oxygen, can form acetic acid, by acting on its own elements, we can understand the observed augmentation in the weight of this acid, and we shall be inclined to attribute its production in a general manner to the transformations undergone by the ferment while it acts upon the sugar. We may say the same thing of the leucine and tyrosine found by Béchamp in the extract of fermented glucose. These are compounds in the production of which the sugar and the fermentable matter take no part.

However, the later works of Béchamp on this question are not favourable to this opinion concerning the acetic acid (Comp. Rend., vol. 75, p. 1036). They tend to establish: 1st. That the contact of air, far from augmenting the production of acetic acid, diminishes it. A fermentation which in carbon dioxide produces from ·25 to ·40 of this body for 100 of sugar, gives only ·1 in contact with air; 2nd. That the acetic acid comes from the sugar, and not from the ferment, for by arranging the experiments suitably, we may obtain a weight of acid greater than that of the ferment employed. In proportion as the ferment is well nourished, and the better it multiplies, the less acetic acid it yields. That which has only cane sugar for its nourishment is exhausted, and produces more acetic acid. The temperature and the augmentation of pressure tend to increase this phenomenon.

Most of the natural saccharine juices, such as those of beet-root and grape-cake, give rise to the production of small quantities of alcohols homologous with ordinary alcohol. We find, in fact, when large masses of products are operated upon, in the arts, and crude alcohol is distilled carefully, by means of suitable rectifying apparatus, that there is a residuum less volatile than ethyl alcohol, having a strong and disagreeable odour. This oily residuum, known under the name of oil of potatoes, has been made the object of many researches by Chancel, Wurtz, Pelletan, Faget, and others; they have generally found it composed in great part of propyl alcohol, $C_3 H_8 O$, butyl alcohol $C_4 H_{10} O$, dominant amyl alcohol, $C_5 H_{12} O$, caproic alcohol, $C_6 H_{14} O$, œnanthyl alcohol $C_7 H_{16} O$, and caprylic alcohol $C_8 H_{18} O$.*

M. Jeanjean has, besides, ascertained, in the products of the distillation of the water in which fermented madder has been washed, the presence of camphyl alcohol (camphor of Borneo, $C_{10} H_{16} O$).

We may ask whether these secondary products, which are relatively not very abundant, owe their origin to alcoholic fermentation properly so called, or to distinct concomitant fermentations, having each a special ferment; or whether, in fact, it is better to attribute their appearance to special principles accompanying glucose in the natural saccharine juices.

* I have followed Schorlemmer in using the terms, ethy*l* alcohol, propy*l* alcohol, instead of ethylic. The change is unimportant.

The actual state of science does not allow us as yet to answer these questions definitively.

M. Berthelot points out (Ch. Org. Fondée sur la Synthèse, vol. 2, p. 631) that the production of all these homologues of ordinary alcohol at the expense of sugar may be formulated by the general equation:—

$$\frac{n}{4}\left\{ C^6 H^{12} O^6 \right\} = C^n H^{2n} + O^2$$
$$+ \frac{n}{2} C O^2 + \frac{n-2}{2} H^2 O.$$

Of the Fermentable Body.

The progress of chemistry has taught us to distinguish many varieties of saccharine hydro-carbons, differing either in their properties or in their composition; they do not all show the same characters, when they are subjected to the influence of the special alcoholic ferment, the yeast of beer.

Glucose (grape sugar or starch sugar) levulose or the sugar of acid fruits, uncrystallizable sugar, maltose or sugar of malt, formed by the action of diastase on dextrin, lactose or sugar derived from sugar of milk (lactive) by the action of acids, all have the same formula, $C_6 H_{12} O_6$.

An almost entire resemblance between their behaviour in presence of a ferment corresponds with this analogy in their composition. These sugars are split up progressively into alcohol and carbon dioxide without undergoing any previous transformation—as Mitscherlich has observed (Ann. de Poggen., vol. 135, p. 95), the rotatory power of a solution of glucose diminishes in proportion to the quantity of alcohol produced.

The equation of Dumas and Gay-Lussac, modified by Pasteur, applies without any restriction to these various saccharine matters. We will only add that, according to the interesting observations of Dubrunfant, glucose mixed with levulose, with the addition of yeast, ferments sooner than the latter does when by itself. This is what always happens when, having altered cane sugar by an acid, we subject to fermentation the mixture of equal weights of glucose and levulose which results from this alteration ; the glucose disappears before the levulose, which, last of all, undergoes alcoholic decomposition. Dubrunfant has given this phenomenon the name of elective fermentation. The sugars whose composition is represented by the formula $C_{12}\,H_{22}\,O_{11}$ can also be fermented, but only on the condition of being previously hydrated, which converts them into sugar with the formula $C_6\,H_{12}\,O_6$.

Saccharose or cane sugar is changed, when hydrated, into two isomeric molecules, one of which crystallizes and causes the plane of polarized light to deviate to the right, and the other remains uncrystallizable, and turns it to the left (lévulose). The two products of this splitting up are fermentable, the hydratation, as is well known is effected under the influence of acids, of water only, of light, and of the lower cellular plants. It may be understood, from this last observation, why cane sugar can ferment : as soon as it is placed in contact with yeast, it begins to alter, and afterwards the glucoses produced form alcohol. The ferment therefore plays a double part towards the saccharine matter, of which the first is much simpler than the second. The alterative power is due to the presence in the ferment of a soluble and inorganic nitrogenous principle, formed at the expense of the proteids of this organism. This active alterative substance accumulates more especially in great proportion in ferment which has been left to itself, and which has undergone the phenomenon called softening.

The water in which such ferment has been washed alters cane sugar with such rapidity that, when we mix the two liquids (sweetened water and the water in which the ferment was washed), the liquid rapidly reduces Fehling's liquid poured into it some seconds after. We shall return to this order of phenomena when we speak of indirect fermentation, called fermentation by a soluble ferment.

Meletizose, melitose, and lactine or sugar of milk, are in the same category as cane sugar, and must be previously hydrated. Berthelot observed this remarkable peculiarity in melitose; only half of this sugar is decomposed into alcohol and carbon dioxide, the other is transformed into a compound, isomeric with glucose, namely, eucalin, which is not fermentable.

All bodies capable of producing glucose and its congeners by hydratation, belong to the class of indirectly fermentable substances, such as starch, dextrine gum, glycogea, and the various glucosides which are found in vegetable tissues.

CHAPTER III.

ALCOHOLIC FERMENTS.*

We have hitherto considered bodies susceptible of alcoholic fermentation, and the details of the reaction known by this name; there remains for us to speak of the most interesting part of our subject, the exciting cause of fermentation. It is more especially on this, the most obscure and the most difficult part of the question, that the most varied, and, we may say, the most lively discussions have been raised. The other parts of the problem required for their solution, in fact, only good analysis and rigorous quantitative experiments.

Historical.—Leuwenhoeck was the first, in 1680, to examine beer-yeast by the microscope, and to ascertain that it is formed of very small spherical or ovoid globules. He could not, however, determine their nature.

In his memoir on fermentation, presented to the Academy of Florence (1787), Fabroni compared ferments to animal substances. "The matter which decomposes sugar is a vegeto-animal substance; it resides in peculiar utricles, in grapes as well as in corn.

* Cfr.—Pasteur, "Annales de Chimie et de Physique," 3rd series, vol. 58, p. 364; Monoyer, "Thèse de Médecine," Strasburg, 1862; L. Engel, "Thèse pour le Doctorat ès Sciences," Paris, 1872.

When the grapes are crushed, this glutinous matter is mixed with the sugar; directly the two substances come in contact, effervescence and fermentation commence."

The experiments and conclusions of Fabroni did not seem sufficiently to elucidate the question, for in the year VIII. the class of physical sciences of the institute proposed as the subject for the prize the following question:—

"What are the characteristics which, in vegetable and animal matters, distinguish those which serve as ferments from those which they cause to undergo fermentation?" Three years afterwards,* Thénard presented a remarkable memoir on alcoholic fermentation and ferments. He arrived at the conclusion that all natural juices, when spontaneous fermentation has set up, give a deposit which has the appearance of beer-yeast, and like it, is able to ferment pure sweetened water. This ferment is of an animal nature; it is nitrogenous, and gives over much ammonia when distilled. When Thénard said that ferment is of an animal nature, he considered only its chemical composition, and made no allusion to the organization of the ferment. We shall return to the labours of this investigator when we come to study the transformations undergone by the ferment during the act of fermentation.

Gay-Lussac proves, by well-known experiments, that fermentation is only developed in the must of grapes when it has been placed for a moment in contact with air; he concludes, from his experiments, that oxygen

* "Ann. de Chimie," vol. 26, p. 247

is necessary to commence the fermentation ; but that it is not required to continue it.

In 1828, Colin made many experiments which seem to show that a great number of organic nitrogenous substances differing from yeast, and in process of change, are able, when placed in sweetened water, to set up alcoholic fermentation at the end of some hours ; at the same time the fœtid odour of putrefaction is changed into the agreeable smell of the must of wine (Colin, Ann. Chim. Phys., 2nd series, vol. 28, p. 128, 1828).

The question of fermentation had reached this point, and yeast was regarded as an immediate principle of plants, having the property of becoming precipitated in presence of fermentable sugars, when Cagniard de Latour took up the incomplete microscopical observations of Leuwenhoeck, which had been so long forgotten.

He observed that yeast is a mass of organic globules, susceptible of reproducing themselves by means of buds, or seminules, which appeared to belong to the vegetable kingdom, and not to be simply organic or chemical matter, as had been supposed. He concluded that it is very probably by some effect of their vegetation that the globules of yeast disengage carbon dioxide from the saccharine liquid, and convert it into spirituous liquor. (Ann. Chim. Phys., 2nd series, vol. 68.)

The discovery of Cagniard de Latour was again made almost at the same time, but independently, by Dr. Schwann at Jena, and by Kützing at Berlin (Schwann, Poggen. Ann., 1837, vol. 41, p. 184; Kützing, Journ. für Prakt. Chem. 2, p. 385); confirmed by the observations of Quevenne (Journal de Pharm. 2, vol. 24), of Turpin (Comp. Rend. de l'Acad., 4, p. 369), of Mitserlich (Poggen. Ann., 55, p. 225), it led, without any possible contradiction, to the following conclusions respecting the nature of yeast. This body was considered to be a mass of organized and living cells, composed, like vegetable or animal cells, of an envelope and granular contents.

From the very commencement, it was not agreed what place should be assigned to this new form of life. Some saw in it a fungus without a mycelium, others looked upon it as one of the algæ.

Thus Turpin (Comp. Rend., p. 379, 1838) placed the cells of yeast in the genus Torula of Persoon, in which "sporæ in floccos moniliformes concatenatæ, dein seudentes." These cells were thus compared to spores, without considering their mode of production, which is quite different, and without remarking that the Torula has a mycelium, which is never found in ferments.

The discovery of true spores has since proved that ferments cannot be placed in the family of the Torulaceæ. Meyen (Pflanzen Physiologie, vol. 3, p. 455) also considered that yeast was a fungus, and created a new genus for it, under the name of Saccharomyces. This name was adopted by Rees, Engel, &c.; Kützing, on the contrary, with many other authors, maintained that ferments are algæ, which he arranged in a separate genus, *crypto-coccus.*

The opinion of men of science who wish to assimilate yeast and ferments in general to algæ, was founded on the observation that these cells multiplied by budding only. But we shall soon see that by placing these ferments under certain conditions, we succeed in causing them to fructify ; besides this, algæ almost always contain chlorophyll, which fungi and ferments do not.

It is now, therefore, very generally admitted that ferments are fungi. Without in the least detracting from the merits of Cagniard de Latour, we ought to say that he had been anticipated in the field of microscopic observations, not only by Leuwenhoeck, but also by Kieser (1814, Schweigger's Journal, 12, p. 229), who describes it as formed of small transparent motionless spherical corpuscules, all of nearly the same size ; also by Desmazières, who, in 1826 (Annales des Sciences Naturelles, vol. 10, p. 4), examined the pellicle formed on the surface of beer, and called by Persoon *mycoderma cerevisiæ.*

Desmazières was the first to give a representation of the globules which he had observed, and having seen in them a particular movement, which is nothing more than the Brownian movement, which at that time was unknown, he arranged these globules among the *animalcula monadina.* Astier, as early as 1813 (Ann. de Chimie, vol. 87, p. 271), did not hesitate to affirm that ferment, recognized as an animal substance by Fabroni, was alive, and derived its nourishment from the sugar, whence resulted the rupture of equilibrium between the elements of this body. By this theory, it is easily explained, said he, that all the causes which kill animals, or hinder their development, must be opposed to fermentation.

The observers who had demonstrated the organic nature of ferments established at the same time that in a great number of liquids in alcoholic fermentation (such as natural saccharine juices, solutions of sugar with albumin, &c), globules of ferment are formed, as Thènard had observed. Schmidt of Dorpat arrived at the same conclusions by repeating Colin's experiments on the fermentation excited by albuminoid matter in the process of decomposition. Microscopic observations revealed to him the development of globules of ferment whenever there was the production of alcohol.

From all these successive observations, which were complementary to each other, arose the opinion generally admitted, that yeast accompanies every well marked alcoholic fermentation; it seems, therefore, that the theory propounded by Astier and Cagniard de Latour concerning fermentation, ought to have prevailed, and to have been received by men of science. But this was not the case. From the very commencement of the discussion, the conclusions of Cagniard de Latour and of Schwann found a powerful opponent. Liebig, whose name was then an authority in chemistry, had a decided theory concerning the phenomena of fermentation in general, and he defended it with vigour, even after the experiments of Pasteur, who admits that alcoholic fermentation is an act connected with life, and with the organization of globules.

"My most decided opinion," says Pasteur, "on the nature of alcoholic fermentation is the following: The chemical act of fermentation is essentially a correlative phenomenon of a vital act, beginning and ending with it. I think that there is never any alcoholic fermentation without there being, at the same time, organization, development, multiplication of globules, or the continued consecutive life of globules already formed." As to the hypotheses which tend to go more deeply into the physiological cause of decomposition, M. Pasteur neither admits nor rejects them, at least in his first memoir. Thus the ideas of Pasteur confirm and extend those of Cagniard de Latour.

As to the theory of Liebig, it does not differ from that of Willis and Stahl. According to the German chemist, the cause of fermentation is the internal molecular motion, which a body in the course of decomposition communicates to other matter in which the elements are connected by a very feeble affinity. "Yeast, and, in general, all animal and vegetable matters in a state of putrefaction, will communicate to other bodies the condition of decomposition in which they are themselves placed; the motion which is given to their own elements by the disturbance of equilibrium is also communicated to the elements of the bodies which come into contact with them. (Liebig, Ann. de Chimie et de Phys., 2nd series, vol. 71, p. 178.) This very philosophical and seducing explanation of an obscure phenomenon obtained greater credit among men of science because it gave the key, not only to alcoholic fermentation, but also to other phenomena of the same kind, such as the transformation of sugar into lactic and butyric acids, in which organic products had not hitherto been observed, and which seemed only to be the results of a conflict between a fermentable substance and a nitrogenous substance in process of putrefaction. Fremy and Boutron supposed that, in matters capable of acting as ferments, the character of the fermentation varies with the degree of decomposition of the substances. This would be successively alcoholic, lactic, or butyric ferment according to the more or less advanced state of its decomposition.

It is thus that the recognized invariable presence of an organic body in every liquid in process of alcoholic fermentation began, by degrees, to be considered as a fact of slight importance with regard to the reaction; the latter is excited, not by the direct action of the globules of ferment, considered as a living organism, but by the decomposition of the proteic nitrogenous matter of this ferment, regarded only as nitrogenous substance. The experiment of Gay-Lussac was naturally interpreted by this opinion; the momentary presence of oxygen was indispensable to set up the molecular disturbance of the albuminous matter of the must of grapes.

Berzelius, for his part, treated the organic nature of yeast as a poetico-scientific reverie, and, rejecting the doctrine of Liebig, borrowed from Willis and Stahl, would only see in fermentation an act of contact due to *catalytic* force, and in yeast an amorphous principle. Mitscherlich supported the ideas of Berzelius, while he admitted the organic nature of the ferment.

However, the clear and well conducted researches of Pasteur on fermentation, and especially on alcoholic fermentation, had partly reconciled men of science to the physiological theory; wherefore Liebig thought right to recommence the contest in favour of his own ideas. In 1870, he published a long memoir on fermentation and the source of muscular force (Ann. der Chemie und Pharmacie, vol. 153, p. 1), a memoir in which he sought to show that the principal experiments of Pasteur are not conclusive. He first demonstrated that the physiological theory of Pasteur, which explains the decomposition of sugar by the nutrition and development of an organic substance, is not opposed to the mechanical doctrine of which he is the champion. This avowal was already an enormous concession made by the German chemist, almost an avowal of defeat, for this language is very different from that which he had formerly used.

However, the attack was sufficiently powerful to induce Pasteur to reply to it (Ann. Chimie Phys., 4th series, vol. 25, p. 145, 1872), and Dumas to undertake a series of experiments, in order to ascertain if it were possible to verify the consequences of Liebig's theory.

Dumas, by means of ingenious experiments, conducted with his never-failing precision (Ann. de Chimie et de Physique, 5th series, vol. 3, p. 69), succeeded in proving irrefutably; 1st. That saccharine liquids are not influenced by ferment, even through the shortest columns of liquid, the thinnest membranes, or even without any separating medium; and that its immediate and direct contact is necessary; 2nd. That sonorous vibrations have no influence on the movements of fermentation; 3rd. That no chemical action, among a great number of those which have been tried, has been able to effect the decomposition of sugar into alcohol and carbon dioxide.

These negative results, without bringing about any decisive solution of the question, are, however, contrary to the opinion of a transmitted movement.

We may thus sum up these three great theories of fermentation: 1st. The vitalist theory, formulated by these words of Turpin, "Fermentation as effect, and vegetation as cause, are two things inseparable in the act of decomposition of sugar," maintained by Astier, Cagniard de Latour, Schwann, Kützing, Turpin, Bouchardot, Van de Brock, Shroeder, Pasteur, and Bichat; 2nd. The mechanical theory of Willis, Stahl, and Liebig, admitted by Gerhardt; 3rd. The theory of catalytic forces, and of acts of contact, maintained by Berzelius and Mitscherlich.

Various mixed opinions range themselves by the side of these three theories. Thus M. Berthelot considers fermentation as produced by the action of a substance elaborated by organic ferments, comparing, with this idea, the alcoholic and lactic fermentations to the conversion of starch into dextrine and sugar under the influence of diastase, a soluble inorganic ferment. The learned chemist supports his opinion by experiments which prove that, in certain cases, there may be the production of alcohol without the formation of ferment (Chimie Organique Fondée sur la Synthèse); which does not exclude the fact, now distinctly established, that fermentation may be excited, and is indeed energetically originated, by special organic substances.

As to a more precise relation between chemical phenomena and the physiological functions of the organic ferment, it is still to be discovered; and all that has been said, written, and brought forward to decide the question needs experimental proof, and can only be considered by us in passing.

No one doubts that, in organic living cells, whether they be isolated, like those of yeast, or form an integral part of a more complicated organism, there resides a special force, capable of producing chemical reactions under conditions quite different from those which we employ in our laboratories, and to produce results of the same class. This force, which we imagine to be as material as heat, reveals to us its activity by decompositions effected on complex molecules. Whether we reduce the problem to the action of a soluble product elaborated by the organic ferment, and to which it has communicated its power, or suppose that the whole of the ferment exercises an action of this kind, we ultimately arrive at a motion communicated, more or less directly, by vital force, and dependent upon it.

We must not confound this interpretation of the phenomena with Liebig's theory; the German chemist held that the albuminoid principles, in decomposing spontaneously, produce the molecular motion which is transmitted to the sugar to split it up; here, on the contrary, it is the living organism which develops force, by borrowing it from the great external reservoir and transforming it. This force may act chemically, either directly or indirectly by means of a soluble ferment.

Description of the Ferment.—We will give, with Rees, the name *Saccharomyces cerevisiæ*, to the alcoholic ferment of beer; it is the ferment which has been most thoroughly studied, and which is the most easily procured.

There are three methods of causing the wort of beer to ferment; the two first are generally employed; these are surface and sedimentary fermentations.

In the third method, employed in Belgium only, the wort is left to itself in a locality situated above the level of the ground, and the spontaneous development of fermentation is waited for; in the two former instances, the action is excited by mixing with the wort a suitable proportion of yeast arising from an anterior operation of the same class.

In beer brewed by surface fermentation, the starch of the malt is changed into sugar by its being steeped several times successively; fermentation takes place in casks at a comparatively high temperature, from 15° to 18° C. (59° to about 65° Fahr.) The yeast, in this case, as it is formed, rises through the bung-holes, at the upper part of the cask. In England, this fermentation is carried on in large open vats; the yeast then floats on the surface of the liquor, and can be skimmed off.

In the manufacture of beer brewed by sedimentary fermentation, the saccharification is effected by de-

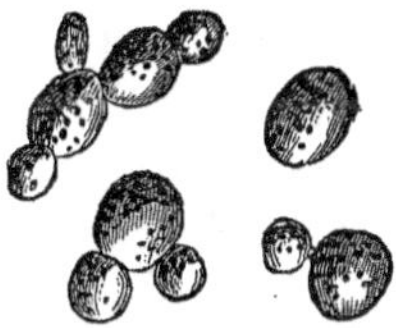

FIG. 1.—Saccharomyces cerevisiæ—Yeast of sedimentary beer, × 400.

coction, and the transformation takes place in open vats, at a temperature not exceeding from 12° to 14° C. (from about 53° to 58° Fahr.) The yeast is deposited at the bottom of the vats, and adheres in the form of a pasty mass. When once the first and most active fermentation is over (it usually lasts two or three days for the surface fermentation, and eight or ten days for the sedimentary), the clear liquid is drawn off, and kept in casks, glass, or stone bottles. In the meantime the separation of the yeast has not been completed; it continues to act on the still unmodified sugar; therefore the amount of alcohol and carbon dioxide yielded increases with the time of keeping, and at the same time the liquid becomes turbid by the production of fresh yeast.

The yeast greatly exceeds (seven or eight times), in weight and in volume, that which had been previously introduced into the wort. To this fact, which we now merely notice in passing, we shall return presently; it is explained by the multiplication by buds, which takes place whenever the cells of yeast are placed in a saccharine medium suitable to their development; and the wort of beer affords excellent conditions in this respect. After the sedimentary fermentation, the yeast found at the bottom of the vat is composed, almost entirely, of cells of a single species of alcoholic ferment, the *Saccharomyces cerevisiæ* (Fig. 1). The microscope will show, but in a very small proportion, granules of lupulin, crystals of calcium oxalate, spores and mould. This deposit is of the consistence of a paste of a yellowish white, or yellow-ochre colour.

The cells are round or oval, from $\frac{8}{1000}$ to $\frac{9}{1000}$ of a millimètre (from ·00031 to ·00035 in.) in their greatest diameter. They are formed of a thin and elastic membrane of colourless cellulose, and of a protoplasm, also colourless, sometimes homogeneous, sometimes composed of small granulations. We find in the protoplasm one or two vacuoles, of various sizes, containing cellular juice. The cells are either separate, or united two by two.

When these cells are deposited in a fermentable liquid, we soon see at one, and more rarely, at two points of their surface, vesicular prominences arise, the interior of which is filled at the expense of the protoplasm of the mother-cell; these prominences enlarge, and at last, having attained the size of the original cell, they lessen in diameter at the base (Fig. 2). They usually originate at the widest side, but more rarely at the extremities. As soon as the formation of this kind of neck takes place, the new cells separate with considerable rapidity from the mother-cell, in which the protoplasm, given up to the young cell, is replaced by one or two vacuoles. If the conditions are favourable, the same cell is able to produce several generations of cells; but, by degrees, it loses all its protoplasm, which at last unites in granules swimming in the midst of superabundant cellular juice. The cell then ceases to reproduce, and even to live; the membrane is ruptured, and the granular contents are diffused in the liquid.

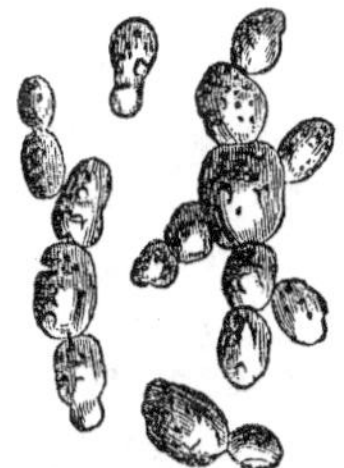

Fig. 2. — Saccharomyces cerevisiæ— Yeast of sedimentary beer, budding, × 400.

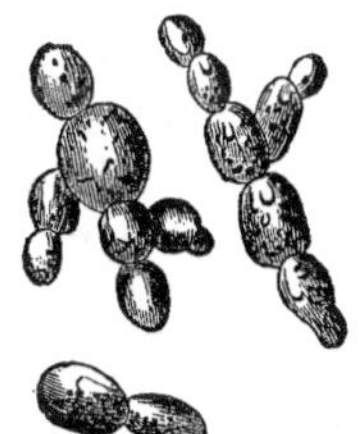

Fig. 3. — Saccharomyces cerevisiæ— Yeast of surface beer, budding, × 400.

When the *Saccharomyces cerevisiæ* is not in contact

with a fermentable liquid, it may remain for some time without becoming modified.

The isolated cells of the surface ferment (Fig. 3) do not differ sensibly from those of the sedimentary ferment; and although it has been maintained that the larger and oval forms are more prevalent in it, it is difficult to establish any distinction of this kind, for we find in the two varieties all the intermediate forms between the two extremes.

Besides, an elevation of temperature above 14° C. (57·2 Fahr.) during the fermentation is sufficient to augment considerably the size of the sedimentary cells, to cause them to attain a large diameter, from $\frac{13}{1000}$ to $\frac{14}{1000}$ of a millimètre (·00051 to ·00055 of an in.), giving them a long oval form; at the same time, we see two circular vacuoles make their appearance; one, large, situated near the larger end; the other, smaller, is found in the narrow part of the cell.

The surface *Saccharomyces* buds much more quickly than the other variety, when placed in a fermentable liquid (Fig. 3). This budding is very rapid; the different cells which issue from each other remain attached together, forming small ramified chains, composed of from six to twelve, and even more, individual buds. It may be easily understood that the bubbles of gas adhering to these chaplets have greater hold upon them than on an isolated cell; this causes the newly-formed yeast to be raised towards the surface of the liquid, and this is effected the more rapidly when the fermentation is more active. In these chaplets, the cells have an elliptical form. The only well-ascertained difference between the two kinds of yeast is, therefore, the rapidity with which buds are formed, and the greater activity of

Fig. 4.—Saccharomyces cerevisiæ—Surface yeast, at rest, × 400.

Fig. 5.—Saccharomyces cerevisiæ—Sedimentary yeast, in a growing state, × 400.

the ferment in the one case; but this does not authorize us to consider these two ferments as belonging to different species. We are able, indeed, though with great difficulty, by changing their conditions of existence, to transform one into the other.

Multiplication by buds is not the only mode of reproduction of the *Saccharomyces cerevisiæ*. It is true that it alone appears as long as the yeast is in contact with an appropriate fermentable liquid. We owe to Rees (Botanische Zeitung, December, 1869) the discovery of the fructification of yeast, and of ferments in general; that is to say, their reproduction by means of spores. As to the *Saccharomyces*, the conditions which appear to be peculiarly favourable to this special evolution of the fungus, are to deprive it suddenly of all nourishment, especially saccharine, and to expose it to a damp atmosphere, or, still better, to place it on a substance capable of affording it sufficient and constant humidity.

We obtain, according to Rees, the richest production of spores by leaving yeast previously washed several times, for some days in contact with distilled water, and then, having decanted the greater part of the water, later on removing day by day the small portions of water which become separated from it. Under favourable conditions, we obtain, at the end of fifteen or sixteen days, a very rich formation of spores; but very often this result is prevented by the putrefaction of the yeast.

M. Engel, from whose essay we borrow the greater part of these details, has also studied this question: he proposes to cause the yeast to fructify, by the following contrivance:—

We mix up some plaster, and allow it to run on some polished, but not oily surface, such as window-glass, plate-glass, or marble. We make up the mass into some form corresponding with the interior of the vessel in which we intend to preserve it. The dimensions in each direction ought to be about two centimètres (·78 in.) smaller than the internal dimensions of the vessel, so as to allow a space between the sides and the mass of plaster sufficient to pour in some distilled water. We then take some very fresh yeast, decant as far as possible all the supernatant fermentable liquid, and dilute the yeast with distilled water, so as to bring it to the consistence of very fluid broth; we pour some drops of this liquid on the polished surface of the plaster, inclining the mass in every direction so as to spread the solution uniformly. This operation should be performed quickly, for the plaster absorbing the water very rapidly, the diluted yeast would become too thick, would not spread with sufficient uniformity, and the layer of ferment would be too thick in certain parts. We then place the mass in the vessel, with the part covered with yeast upwards, and pour, by means of a funnel, distilled water between the sides of the vessel and the piece of plaster, until the liquid reaches above a centimètre ('39 of an in.) beneath its upper surface. The vessel is then covered with a plate of glass to prevent, as far as possible, the contact of dust, and of spores floating in the atmosphere.

Under these conditions, the vegetative life of the yeast ceases suddenly, and in a few hours we see great changes take place in the protoplasm of the cells. The oldest, and those which are least rich in protoplasm, perish and break up. Others, on the contrary, grow larger, their lacunæ disappear, and the protoplasm is diffused uniformly in the cellular juice. At the expiration of from six to ten hours, we notice the appearance, in the midst of the protoplasm, of from two to four small "islets," more brilliant and dense than the rest, around which fine granulations collect. These dense spots do not present any appearance of a nucleus, and they become differentiated more and more till they are

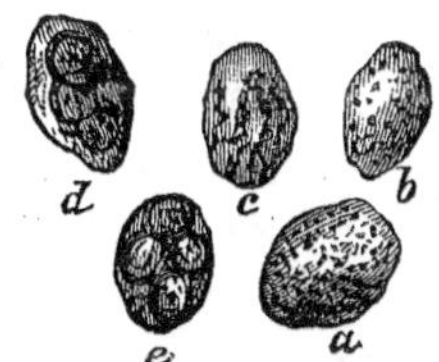

FIG. 6.—Saccharomyces cerevisiæ—Formation of spores, × 750. *a, b, c, d, e*, successive phases of the production of spores.

FIG. 7.—Triads of spores, germinating, × 750.

exactly spherical (Figs. 6 and 7). Twelve or twenty-four hours later, each of these spherules is invested with a membrane, very thin at first, but which thickens by degrees, and then shows a double outline, when magnified 600 diameters.

The spore is then ripe. The mother-cell thus contains from two to four spores. When there are but two, they are situated in the greater diameter; when there are three, they are usually arranged in a triangle; when there are four, they are in the form of a cross, or three of them form a triangle, on which the fourth is superposed in the manner of a tetrahedron.

During their evolution the spores touch each other; a plane surface is therefore produced at the point of contact; they remain attached to each other during some time after maturity, and thus form combinations of two, three, and four. The two spores connected together have only one plane surface, the triads have two, inclined to each other at about 120°, and in the tetrads arranged in the form of a cross we observe, also, two plane surfaces at right angles. When the spores ripen, the thecæ are moulded on them, and thus assume their various forms. The theca of the diads is elliptical; that of the triads is triangular, with rounded angles; that of the tetrads, in the shape of a cross, is in the form of a diamond with rounded angles; in the tetrads, piled up on each other, the theca is tetrahedral; when in complete maturity, the membrance of the spore case, or mother-cell, which has become a fruit, is torn, and allows the spores to escape. The thecæ themselves vary from $\frac{10}{1000}$ to $\frac{15}{1000}$ of a millimètre (·00039 to ·00047 of an in.), and the spores from $\frac{4}{1000}$ to $\frac{5}{1000}$ of a millimètre (·00015 to ·00019 in.)

The innumerable quantity of thecæ which we obtain by the method of fructification on plaster, leaves no doubt as to the origin of these organisms; besides, Rees and Engel have often observed thecæ still attached to vegetative cells, and those in process of budding, and have thus recognized their relationship.

Hitherto, all that we have said applies to sedimentary beer, properly so called, to the *Saccharomyces cerevisiæ;* but this special fungus is not the only one capable of determining the alcoholic fermentation of glucose. Microscopists who have studied this difficult question distinguish, besides the special ferment of beer, other varieties, for the most part belonging to the genus *Saccharomyces* of Meyen.*

Under the name of *Saccharomyces minor*, Engel describes a kind of ferment obtained by him from the leaven of flour, and to which it owes its activity.

The process of extraction is similar to that employed by chemists to separate starch from gluten in flour. The liquid which passes through the sieve when bakers' yeast is kneaded under a slight stream of water, contains starch and globules of yeast, which, on account of their smaller density, are deposited last. We may thus, by a series of washings, obtain a product very rich in globules of yeast, and poor in grains of starch. Engel proposes to employ sweetened instead of pure water in these washings, in order not to diminish the physiological activity of the cells.

* Simple thecaphorous fungi, without a true mycelium. The vegetative organs are cells, produced usually by buds from similar cells, and which, detaching themselves, sooner or later, from the mother-cell, multiply in the same manner. A part of the cells thus formed are transformed, in another medium, into naked sporiferous thecæ; unicellular spores, to the number of from one to four in each theca. The germination of spores reproduces directly vegetative cells analogous to those which originate from buds.—ENGEL, *Thèsis of the Faculty of Sciences of Paris*, 1872.

The ferment, examined by the microscope, is seen under the form of isolated globules, double or sometimes united in threes. The largest of these globules are $\frac{6}{1000}$ of a millimètre (about ·000315 of an in.) in diameter; their vacuoles are less apparent than those of the yeast of beer.

This ferment, sown in the most favourable saccharine medium, prepared according to the formula of M. Pasteur, has only produced a very slow fermentation. By renewing the experiment seven times, and only making use each time of the ferment obtained in the preceding experiment, we see no apparent modification in the form and dimensions of the globules. The budding of this species is effected in the same manner as in the yeast of beer.

Placed under the conditions favourable to the formation of spores, of which we have spoken before, the *Saccharomyces minor* is transformed into sporiferous thecæ of spherical, and occasionally though rarely of ovoid form, of $\frac{6}{1000}$ to $\frac{7}{1000}$ of a millimètre in diameter. The spores are only $\frac{3}{1000}$ of a millimètre in diameter, and are united in diads or triads. In fact, except their form, which is always spherical, their smaller dimensions and greater activity, the ferments of bread resemble that of beer.

The *Saccharomyces ellipsoïdeus* of Rees is nothing but Pasteur's ordinary alcoholic ferment of wine (Études sur le Vin, Figs. 8, 9, and 11); it ought not to be confounded with the *Cryptococcus vini* of Kützing. The adult cells have an ellipsoidal form $\frac{6}{1000}$ of a millimètre in length by $\frac{4}{1000}$ or $\frac{}{1}{}^{5}$ in breadth (·00024 by about ·000176 in.), with an oval vacuole. The sporulation and budding differ in no respect from the analogous phenomena which are observed in yeast (Figs. 8, 9, and 10).

Rees gave the name of *Saccharomyces Pastorianus* (Fig. 11), to a variety of alcoholic ferment of wine observed by Pasteur (Fig. 7, Études sur le Vin). The cells are oval, pyriform, or elongated like a club. The ovoid cells are $\frac{6}{1000}$ of a millimètre in length (·000236 in.), those that are club-shaped, which are seen to proceed like buds from ovoid cells, reach $\frac{18}{1000}$ or

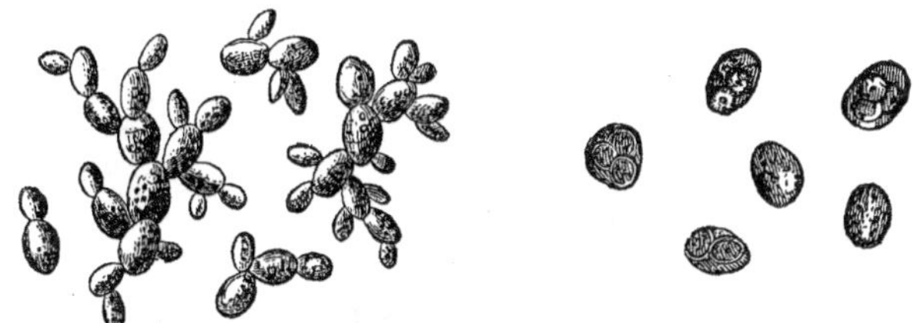

FIG. 8.—Saccharomyces ellipsoïdeus, in process of budding, × 600.

FIG. 9.—Saccharomyces ellipsoïdeus, development of spores, × 400.

$\frac{20}{1000}$ of a millimètre in length, by $\frac{8}{1000}$ or $\frac{10}{1000}$ in breadth at the larger end (·00064 in. to ·00078 in. by ·000314 in. to ·000397 in.); they are united in flakes containing from three to seven articulations.

FIG. 10.

The *Saccharomyces exiguus* (Rees), Fig. 12, is met

with, like the preceding, in the juices of fermented fruits. The cells are only $\frac{3}{1000}$ of a millimètre in length, by $\frac{2 \cdot 5}{1000}$ in width at the larger end (·000118 × ·000098 in.); it multiplies by budding and sporulation, like all the other varieties of this species.

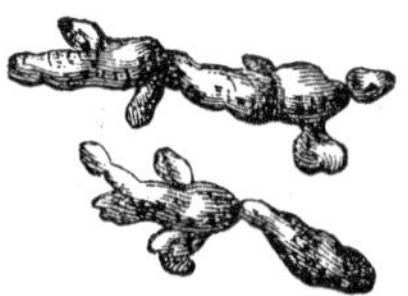

FIG. 11.

The *Saccharomyces conglomeratus* of Rees (Fig. 13), is rather rare; it is met with in the must of wine towards the end of the fermentation. It has spheroidal cells of $\frac{6}{1000}$ millimètre in diameter (·000236 in.).

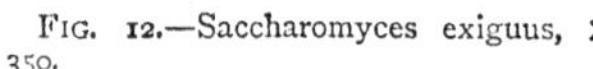

FIG. 12.—Saccharomyces exiguus, × 350.

FIG. 13.—Saccharomyces conglomeratus, × 600.

When the first cell has budded, this bud attains the size of the mother-cell, without being detached; there originate first in the inner angle formed by two cells, and then on different parts of their surface, a considerable number of new cells, which, instead of forming a chaplet or flakes, are entirely conglomerated.

The *apiculated ferment* (*Carpozyma*) does not belong to the genus *Saccharomyces*, according to the observations of M. Engel. This is the most abundant alcoholic ferment. It is met with on all kinds of fruit, especially on berries and stone-fruits, as well as in the greater number of musts of wines in process of fermentation. It has also been noticed in certain kinds of beer, as that of Belgium, and of Obernai; those of Strasbourg do

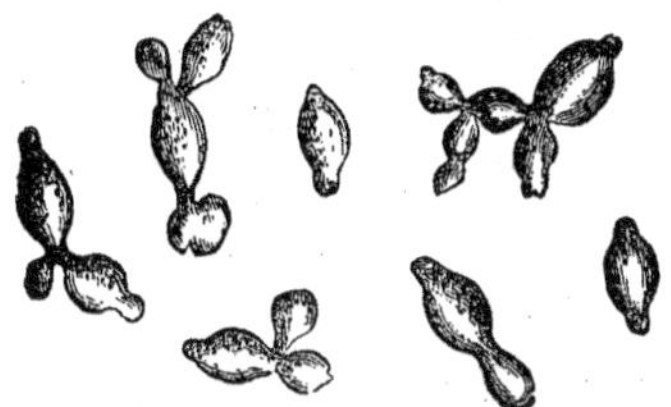

FIG. 14.—Saccharomyces apiculatus (Rees). Carpozyma apic. (Engel), apiculated ferment, × 600.

not contain it. It is this which is usually the first to appear and bud in these musts.

The adult and isolated cells (Fig. 14) have the form of an ellipsoid, whose greater diameter is $\frac{6}{1000}$ of a millimètre (·000236 in.), and the transverse diameter one-half of the larger one. At each extremity there is a small projection or minute stalk, which give the cell the appearance of a lemon. The interior encloses a spherical or ellipsoidal vacuole, around which there is a thin layer of protoplasm, fringed towards the projecting parts. The budding cells always present themselves at the extremity of the projections.

When the development is normal, the new cells extend in the direction of the principal axis of the mother-cell, so that the three cells form a longitudinal row; but when they have ceased to grow, they assume an elliptical form, and bend back at the point of their insertion, so that their longer axis forms at last a right angle with the axis of the mother-cell; one of the cells turns to the right, and the other to the left. They are then detached; at this time they much resemble the cells of *Saccharomyces ellipsoïdeus;* but we soon see the characteristic stalks appear.

When the apiculate ferment is deposited on damp plaster, the transformation into thecæ or sporanges assumes phases very different from those which are observed in the various species of *Saccharomyces*, and resembles the evolutions of the *Protomyces macrosporus* studied by De Bary.

According to Engel, the apiculate ferment is a *Protomyces* without a mycelium; this botanist proposes to give it the name of *Carpozyma.* Its thecæ are spherical, covered with a peritheca, and are hibernating. The development of the spores is very slow, and the spores numerous.

Rees met with a special form of ferment, which accompanies the *Saccharomyces ellipsoïdeus*, in the fermented musts of red wine. It is composed of elongated cylindrical cells. Although this ferment has not been observed in the different evolutions of its vegetative life, it has been thought necessary to establish for it a special species, under the name of *Saccharomyces Reesii.*

The *Saccharomyces mycoderma* (Figs. 16 and 17), called flowers of wine, or flowers of beer, ought also, according to M. Pasteur, to be placed among alcoholic ferments. In fact, although it does not act in this manner, under the ordinary circumstances of its development, when it grows on the surface of fermented liquors, perhaps because the alcohol which it may then produce is destroyed by a subsequent oxidation, M. Pasteur has shown that the *Mycoderma vini*, sown in sweetened water, is able to set up in it alcoholic fermentation.

FIG. 15.—Saccharomyces Reesii, ferment of red wine, × 350.

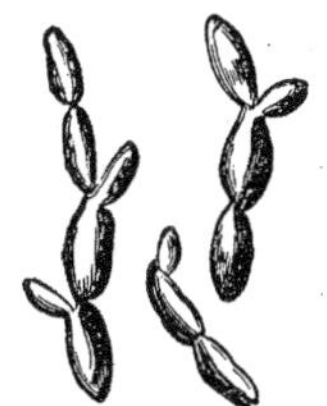

FIG. 16.—Saccharomyces mycoderma, × 350.

It makes its appearance in all alcoholic liquids exposed to the air, when the fermentation is over or has become languid. It grows with great rapidity; it is sufficient to place a few of its cells on the surface of a liquid that easily becomes alcoholic, and we shall find, in less than forty-eight hours, the surface covered with

FIG. 17.—Saccharomyces mycoderma.

a thin pellicle, of a whitish or yellowish tint, at first smooth, and then wrinkled.

M. Engel estimated, by calculation founded on his observations, that in forty-eight hours a cell of *Mycoderma vini* would produce about 35,378 cells. The cells of this *Mycoderma* have various forms—ovoid, ellipsoidal, and cylindrical, with rounded extremities.

The ovoid cells have their greater diameter about $\frac{6}{1000}$ of a millimètre, and their smaller one $\frac{4}{1000}$ (about ·000236 × ·000157 in.). The cylinders have the longer diameter about $\frac{12}{1000}$ or $\frac{13}{1000}$ of a millimètre, and the lesser one $\frac{3}{1000}$ millimètre (·00047 × ·000118 in.).

The cells are generally poor in protoplasm; they show in their interior from one to three brilliant points of fatty matter. The budding is effected at the extremity, by one or two buds originating at each end. Chaplets and ramified and interlaced flakes are thus formed, giving to the whole the appearance of a fine membrane.

When we dilute with a considerable proportion of water the alcoholic liquid on which the mycoderma

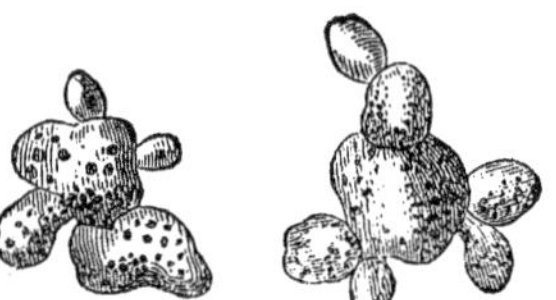

Fig. 18.—Mucor racemosus, ferment in mass.

vegetates, the cells undergo great modifications. The oldest are weakened and die, allowing their protoplasm to escape. The others lengthen, acquiring a larger diameter of from $\frac{16}{1000}$ to $\frac{20}{1000}$ of a millimètre (·000629 to ·000787 in.); their protoplasm collects in different points and forms spores. These are usually about three or four in number, arranged in a longitudinal file in the cell. The spores are generally about $\frac{3}{1000}$ millimètre in diameter (·000118 in.).

Many authors suppose the *Saccharomyces mycoderma* to be derived from the *Sacch. cerevisiæ*, one of the aerial forms of which it represents. This question does not seem yet definitely settled, although there is most reason to suppose that it constitutes a separate and independent species.

We shall have to return to the oxydizing properties of the *Mycoderma vini* when we treat of acetic fermentation.

The *Mucor mucedo* and the *Mucor racemosus* (Fig. 18) possess the property, when immersed in a solution of sugar, and protected from the access of oxygen, of transforming or dividing their mycelium into joints having the form of balls. These balls are multiplied by budding, and excite alcoholic fermentation in sugar as long as they are placed under these abnormal conditions.

This fact, which is indisputably proved, gives considerable support to the theories brought forward by some men of science as to the transformation of ferments, from one to another, according to the conditions under which they are placed. We see, in fact, the *Mucor racemosus* completely change its mode of reproduction when it is placed, without access of oxygen, in a saccharine medium. Analogous facts are known to be produced in the case of other organisms.

These various kinds of ferments have been found, not only in the must derived from fruits, but also on the surface of their pericarps, to which they remain fixed in a state of repose, until by the concurrence of suitable circumstances, they are placed in contact with the saccharine liquid contained in the cells. From this moment they begin to develop by buds, and set up, at the same time, alcoholic fermentation.

According to a recent work by Dr. de Vauréal, the

alcoholic ferment, with its envelope composed of non-contractile cellulose, and reproducing by gemmation, which has been generally believed in, is inadmissible. The supposed budding is only an optical delusion. The utricle of yeast is allied to the spermogones of Tulasue; the granulations or nucleolar elements are spermatia; these elements when set free by the rupture of the utricle, produce new ones.

This mode of multiplication explains the facility with which the reproductive elements of yeast can be carried by the air, when we cannot distinguish in air-dust any characteristic globules of yeast.

In their mode of multiplication, ferments resemble somewhat the zoospores of algæ; when they are not too hybrid, like those of cider, which reproduce a penicillium and an aspergillus, they arrange themselves under the law of metagenesis, like the acalephæ. Especially in the genus *Hydrodiction*, we notice a great similarity.

In fact, we see in this zoospores of two sorts; the greater ones (macrogonidia) are true spores; they have a rapid development and direct evolution; the smaller ones (microgonidia) have a slow development; they do not reproduce the plant, but produce in their interior true zoospores. These are young spores, like ferments.

CHAPTER IV.

ACTUAL COMPOSITION OF FERMENTS.

BEFORE commencing this subject, in which we shall have to consider, among other questions, the chemical modifications which take place in ferments under the different conditions in which they may be placed, we ought to give a summary of the results obtained by experimentalists who have devoted themselves to the study of the chemical composition of these organisms.

Much has been done in this respect. Thus Schlossberger has published very careful researches on the actual elementary composition of the two kinds of ferment of beer, freed as far as possible, by washing and decanting, from the impurities which are found in the crude yeast.

This observer found as the mean of two analyses :—

		SURFACE YEAST.		SEDIMENTARY YEAST.
Results ascertained, the ashes having been removed	Carbon	49·9		48·0
	Hydrogen	6·6		6·5
	Nitrogen	12·1	11·6	9·8
	Oxygen	31·4		35·7
	Ashes	2·5		3·5

Messrs. Mitscherlich, Mulder, and Wagner have published independently the following results :—

		SED. YEAST. (Wagner.)	SURF. YEAST. (Mitsch.)	SURF. YEAST. (Mulder.)	SURF. YEAST. (Wagner.)
The ashes not separated.	Carbon	44·4	47·0	50·8	49·8
	Hydrogen	6·0	6·6	7·2	6·8
	Nitrogen	9·2	10·0	11·1	9·2
	Sulphur		0·6		
	Oxygen	35·8			

M. Dumas (Traité de Chimie), finds:—

Carbon	50·6
Hydrogen	7·3
Nitrogen	15·0
Oxygen, Sulphur, Phosphorus	27·1

This result differs from the others by a larger proportion of nitrogen; but we can understand variations in the composition of such a product as yeast, which is constantly undergoing a process of chemical evolution. Thus, the smaller percentage of nitrogen and carbon furnished by analyses of the sedimentary, as compared with that of the surface yeast, is easily explained, if we keep in mind the fact that the former remains much longer in contact with the liquid after the fermentation. Phenomena dependent on spontaneous change of condition may then take place, transforming into soluble principles a portion of the nitrogenous albuminoid products of the protoplasm, and allowing them to escape into the surrounding liquid.

Schlossberger also endeavoured to isolate the various ultimate principles contained in yeast. By treating it with a very weak solution of potash, filtering and neutralizing the liquid by an acid, he obtained a floccose white precipitate, free from ashes, which, when analyzed for its elements, gave the following numbers:—

MEAN OF TWO ANALYSES.	
Carbon	55·5
Hydrogen	7·5
Nitrogen	13·9
Sulphur	0·

The nitrogen is here present in too small a proportion for a normal albuminoid compound. The analysis agrees tolerably well with that made by me of hemi-protein, one of the products of the decomposition of albumin, by dilute sulphuric acid (*see* the chapter on albuminoid substances). This similarity is the more striking, since the hemi-protein is also soluble in dilute alkalis, and precipitated by acids. The precipitate, when well washed, yields no ashes after combustion. On the other hand, it is very probable that the yeast acts on albuminoid substances by splitting them up progressively, of which we shall find proofs farther on.

By saturating the yeast with acetic acid and precipitating the filtered liquor by ammonium carbonate, Mulder obtained a principle nearer in its composition to albumin, and which gives,—

Carbon	53·3
Hydrogen	7·0
Nitrogen	16·0

We may therefore admit the presence of one or more albuminoid substances in the yeast-cell; in this respect it does not differ from other vegetable cells.

The residuum insoluble in potass, in Schlossberger's

experiment, was then saturated with acetic acid and water. It then shows that it has a composition allied to that of cellulose:—

Carbon	44·9
Hydrogen	6·7
Nitrogen	0·5
Remaining ashes	1·1

This cellulose, boiled with sulphuric acid, is easily converted into fermentable sugar. According to Liebig, it is not soluble in ammoniacal cupric oxide. It seems, therefore, to differ from normal cellulose, soluble in ammoniacal cupric oxide, and which dilute acids do not transform into sugar.

Payen (Mémoire des Savants Étrangers, vol. 9, p. 32) gives the following direct analysis for yeast:—

Nitrogenous matter	62·73
Cellulose (envelopes)	22·37
Fatty matter	2·10
Mineral „	5·80

Other experimentalists, such as Pasteur and Liebig, employing the methods of separation usual in the analysis of the higher plants, have found only 18·5 per cent. of pure cellulose in fresh yeast. If we admit that the nitrogen (11·8 to 12·5 per cent.) contained in the yeast, forms an integral part of the albuminoid matter, we can caculate, by simple rule of three, that the yeast contains about 60 per cent. of proteids and nearly 40 per cent. of hydrocarbons.

This method of calculating the actual composition of yeast is not quite legitimate. We find, in fact, in the washings of fresh yeast, performed in ice-cold water, noticeable quantities of tyrosine, leucine, &c., which contain less nitrogen than albuminoid matters (10 and 7·7 per cent., instead of from 15·5 to 16).

As direct analyses give only 18·5, or at most 30, per cent. of cellulose, one is led to suppose that in the yeast-cell other hydrocarbons are found, more easily attacked by acids and alkalis than is true cellulose. This opinion is corroborated by the production of alcohol during the digestion of yeast, without the addition of sugar, although one cannot discover in it the presence of glucose. On the other hand, we find in the extract of spontaneously decomposed yeast noticeable quantities of a special gummy substance (Béchamp, Schutzenberger). The origin of this gum, if it does not pre-exist fully formed in the fresh cell, can only be attributed to the splitting up of a compound of the family of glucosides, or to a molecular transformation of an insoluble hydrocarbon substance, different from cellulose. Pasteur (Compt. Rend., vol. 48, p. 640) obtained 20 per cent. of sugar by boiling yeast with dilute sulphuric acid.

We owe to Mitscherlich (Ann. der Chemie und Phar macie, vol. 56) some excellent analyses of the ashes of yeast. The dried matter, placed in a silver crucible, itself placed in one platinum, was burnt in a glass tube, in a current of oxygen; the distillation of the organic matter was commenced in an atmosphere of carbon dioxide, and the combustion was finished in oxygen.

Quantity of ash of the yeast:—

SURF. YEAST.	SED. YEAST.	SURF. YEAST.	SED. YEAST.
(Wagner and Schlossberger.)	(Schlossberger.)	(Bull.)	(Mitsch.)
2·5 per cent.	3·5 to 4 per cent.	8·9 per cent.	7·7 per cent.

SED. YEAST. (Mitsch.)	SED. YEAST. (Wagner.)
7·5 per cent.	5·3 per cent.

Taking only the results obtained by Mitscherlich, which appear the most reliable, there would be no difference with respect to the quantities of mineral matters between the two kinds of yeast. The small proportion of mineral matter found by Schlossberger may have resulted from the fact that this experimentalist washed his yeast, an operation which, as Béchamp has shown, gives rise to a continuous elimination of phosphates.

The ash of the ferment gives the following percentage:—

	SURF. FERM. (Mitscherlich.)	SED. FERM. (Mitscherlich.)	SURF. FERM. OF PALE ALE (Bull.)
Phosphoric acid	53·9	59·4	54·7
Potass	39·8	28·3	35·2
Soda	—	—	0·5
Magnesia	6·0	8·1	4·1
Lime	1·0	4·3	4·5
Silica	traces		—
Iron Oxide	—	—	0·6
Sulphuric acid	—	—	—
Hydrochloric acid	—	—	0·1

The predominant elements are, therefore, phosphoric acid and potass, together with a little magnesia and lime.

The analyses of Mitscherlich may be thus calculated:—

	SURF. YEAST.	SED. YEAST.
Phosphoric acid	41·8	39·5
Potassa	39·8	28·3
Soda	—	—
Magnesium phosphate, ($Mg._3\ 2\ PO_4$)	16·8	22·6
Calcium phosphate ($Ca_3\ 2\ PO_4$)	2·3	9·7

The estimates which we have just given of the direct and elementary composition of the yeast are still, as we see, incomplete; and this question is worth attentively studying again.

This opinion has already been brought forward by Pasteur, who thus expresses himself (Ann. Chim. Phys. (3) 58, p. 403):—

"Yeast contains several nitrogenous substances, and also some not nitrogenous—substances distinct from each other. It would be interesting to study this subject. I have found that we should arrive at useful results by examining separately the action of water, of dilute sulphuric acid, and of potass. I think that an examination of yeast, made for this purpose, and of the different materials which compose it, might reveal the secrets of certain changes which are observed in the nature of the extract of the fermented liquid."

Notwithstanding the insufficiency of the data of this question, we see that, considered qualitatively, the cells of yeast resemble other cells which enter into the composition of larger plants, and of fungi in general; they have envelopes formed of cellulose in different degrees of evolution; and their contents are composed principally of an albuminous protoplasm, of hydrocarbons analogous to gum, with fatty and even resinous substances. Indeed, yeast contains a small quantity of bitter resin soluble in alkalis.

When examined quantitatively, it presents a special character, and is very rich in nitrogen, far more so than the vegetable tissues in general. Thus in fungi, Messrs. Schlossberger and Doepping found in 100 grammes of dry matter the following quantities of nitrogen:—

	CANTHARELLUS (CHANTARELLE).	RUSSULA.	LACTARIUS DELICIOSUS.
	gram.	gram.	gram.
Nitrogen, per cent.	3·22	4·25	4·68

	BOLETUS EDULIS.	AGARICUS CAMPESTRIS.
	gram.	gram.
Nitrogen, per cent.	4·7	7·26

The composition of various fungi, according to Payen, is the following:—

	CULTIVATED MUSHROOMS.	MORELLS.	WHITE TRUFFLES.	BLACK TRUFFLES.
Water	91·01	90	72·34	72
Nitrogenous compounds, with a trace of sulphur	4·68	4·4	9·96	8·76
Fatty matter	0·40	0·56	0·44	0·56
Cellulose, dextrin, sugars, tertiary matter	3·45	3·68	15·161	7·59
Salts, Alkalne, calcic, magnesic, silicic phosphates and chlorides	0·46	1·36	2·10	2·07
Nitrogen per cent. of the dry matter	7·33	—	1·532	1·35

Cultivated mushrooms are, therefore, almost as rich in albuminoid matter as yeast, since 100 parts of dry matter contain 52 parts of these substances. Yeast contains 60.

CHAPTER V.

FUNCTIONS OF YEAST.

YEAST is a living organism belonging to the family of fungi, genus *Saccharomyces*, destitute of mycelium, capable of reproduction, like all the elementary fungi, by buds and spores; its composition, as we have just seen, singularly resembles that of other vegetable tissues, and especially of the plants of the same family. The examination of its biological functions, studied more particularly in their chemical aspect, shows us clearly that this elementary form of life does not differ in essentials from other elementary cells, unprovided with chlorophyll, whether isolated or in groups, and belonging to the more complex organs. It breathes, transforms and modifies its proximate principles in a continuous manner, and certainly in the same way as other cells; like these, it can be multiplied by buds and spores. The only important and decidedly distinctive character which seems to render it a form of life absolutely apart from other forms in creation, was removed from it by M. Lechartier and M. Bellamy, when these chemists succeeded in establishing that the cells of fruits, seeds, and leaves, and even animal cells, are capable of changing sugar into alcohol and carbon dioxide.

From this time, therefore, the accurate study of the

biological functions of yeast no longer appears to us as an isolated chapter in the midst of those which compose general physiology, but, on the contrary, as a particular instance, from which we may draw important conclusions as to the chemical phenomena of living organisms in general. Yeast offers this immense advantage to the observer, that it allows him to make all kinds of experiments on it with great facility. It is like clay, which can be moulded at will, composed of one and the same kind of elementary cells, which enables us to avoid the complications due to the intervention of complicated phenomena.

Normal Conditions of the Life of Yeast.—The condition which we shall call normal in the life-history of yeast are those in which this form of life develops itself, and increases with the greatest activity and energy. They are of two orders, physical and chemical.

With respect to physical conditions, we have only to notice the temperature. The temperature most favourable to the nutrition of yeast is also that which is found advantageous to other cellular vegetable organs; between 25° C. and 35° C. (77° and 95° F.). Above and below these limits, the vital manifestations do not cease until we descend below 9° C. (49·6° F.) or rise above 60° C. (140° F.), the temperature at which albuminoid principles begin to coagulate.

As to the chemical conditions, and the most favourable composition of the medium in which this organism is to live and multiply, they are not so simple.

We owe to the learned and patient labours of Pasteur on this question the best information which we possess on this part of the subject. He has been followed in this interesting investigation by some of his pupils (Duclaux, Raulin), of whose researches we shall have to speak further on.

We can see, *à priori*, that the most favourable medium is that which contains the most appropriate nutritive elements. These elements ought to be those which we find in the organism of the cell, or, at least, principles susceptible of allowing the cell to form with them by synthesis, its immediate component parts. Thus we have seen that yeast contains water in greater or less proportion, mineral salts, especially potassium, magnesium, and calcium phosphates. Water and the alkaline and alkaline-earthy phosphates will therefore necessarily form an integral part of the nutritive medium. We find, besides, a great proportion of nitrogenous substances, either albuminous or otherwise. The food of yeast ought, therefore, to include nitrogen. A question then presents itself: Under what form ought nitrogen to be supplied to the cell in order to be assimilated? Experiment alone can reply.

We know, by the elegant researches of M. Boussingault, that the higher plants are not able to absorb the free nitrogen of the atmosphere. This observer caused a certain known weight of seeds, the average amount of nitrogen in which he had previously ascertained, to germinate in a calcined silicious soil, and watered them with pure water. When the plant, in spite of these unfavourable conditions, had attained a certain development, he ascertained the total quantity of nitrogen contained in the whole. The weight of this did not exceed the original quantity contained in the seed. Yeast forms no exception to this rule.

Assimilation of the Nitrogen of Nitrates.—Agricultural

experiments made on a large scale, and repeated for successive years, as well as those made in chemical laboratories, have established the efficacy of nitrates, and salts of ammonium, among the nitrogenous aliments utilized by plants in general.

The importance of saltpetre as a manure, has been long known; however, with respect to biological phenomena, it is a question whether the nitrate introduced into the soil does not, before it reaches the plant which makes use of nitrogen, undergo changes which reduce it to another form, that, for instance, of ammoniacal salts.

M. Boussingault has given a complete solution to this question by experimenting on plants sown in a medium or a soil absolutely deprived of organic matter, and in which the reduction of nitrates is impossible. The following table gives an idea of the influence of nitrates on vegetation in a barren soil:—

	Weight of the dry produce, the seed being 1.	Vegetable matter elaborated.	Carbon dioxide decomposed in 24 hours.	Acquired by the plants during 86 days of vegetation.	
				Carbon.	Nitrogen.
			grammes.	grammes.	grammes.
When the soil had received nothing . . .	3·6	0·285	2·45	0·144	0·0023
When the soil had received phosphates, ashes, and potassium nitrate .	198·3	21·111	182·00	8·446	0·1666
When the soil had received phosphates, ashes, and potassium bicarbonate.	4·6	0·391	3·42	0·156	0·0027

These numbers show such important differences between the results obtained in the same time, with or without the use of nitrates, all things besides being equal, that an error of interpretation is impossible.

We may then say: In the larger plants the soluble nitrate penetrates into the organism in a state of solution, and there it undergoes changes which finally bring out its nitrogen, under the form of albuminoid matter.

Is it the same with yeast? Can it elaborate its proteic matter by decomposing nitrates?

This question has been discussed. On one side, Dubrunfant (Comp. Rend. de l'Acad., vol. 73, pp. 200, 263) says that he has observed a greater activity as a ferment after the addition of potassium nitrate.

On the other side, Ad. Mayer (Lehrbuch der Gährungs-Chemie, 1874) affirms that he has obtained only negative results in a whole course of researches, directed to this end. Schaer arrived at the same results as Mayer.

This decided difference between the phenomena of the nutrition of yeast and of larger plants is still more remarkable, since simple organisms, much allied to the *Saccharomyces*, act upon nitrates exactly in the same way as terrestial plants of a higher order.

Thus, the mildews which vegetate on the surface of liquids derive considerable nourishment from the dissolved nitrate.

Admitting the observations of Mayer to be correct, we are compelled to interpret these negative results in the following manner:—

In order to assimilate the nitrogen of a nitrate, the

plant must first reduce it. Therefore, either the cell of the *Saccharomyces* does not possess this reducing power over the nitrates which we find in other isolated cells, forming an integral part of more complex organisms, or else the experiments have been made under conditions in which this reducing power has not been able to manifest itself. However this may be, we have not arrived at a final decision respecting the assimilation of the nitrogen of nitrates; and before we determine definitely in the negative it will be advisable to vary the experiments.

Assimilation of the Nitrogen of Ammoniacal Salts.—Agriculturists generally agree in recognizing the efficacy of ammoniacal salts in vegetation. The experiments of Sir H. Davy, Kuhlmann, J. Pierre, Lawes, and Gilbert lead us to the same opinion. It appears probable, according to all the facts, that ammoniacal salts may concur in the nutrition of plants, although it is recognized that, with an equal weight of nitrogen, they act in a less favourable manner than the nitrates. The experiments of M. Bouchardat (Mémoire sur l'Influence des Composés Ammoniacaux sur la Végétation), those of M. Cloëz (Leçons de la Soc. Chim., 1861, p. 167) tend, on the contrary, to establish, 1st, that the solutions of ammoniacal salts, usually employed, do not supply to vegetables the nitrogen which they assimilate; 2ndly, that if solutions at the rate of $\frac{1}{1000}$, and even of $\frac{1}{10000}$, of these salts are absorbed by roots, they act as energetic poisons, and kill the plant rapidly. M. Cloëz cites, on this subject, many instances, which leave no doubt as to the reality of the fact.

If this be the case, we must, in order to make these

apparently discordant results agree, either admit that the ammonia, in order to be absorbed usefully, and without danger to the plant, ought to be presented to it under a special form, perhaps very diluted, or in a peculiar state of combination; or else it must previously undergo, in the soil itself, a transformation into a nitrate. This interpretation, due to M. Cloëz, and which is the reverse of the ideas of M. Kuhlmann (for this observer maintains that, on the contrary, the nitrates employed as manures are only active if they are transformed, in the soil, into ammoniacal salts), is by no means contradictory to what we know of the phenomena of nitrification.

Let us return from the higher forms of plants to ferments, as we have already done before. M. Pasteur was the first to study the influence of ammoniacal salts on the development and nutrition of the *Saccharomyces.*

After having ascertained, by experiments made on a large scale in fermentation, carried on commercially, that the ammonia of the ammoniacal salts contained in the juices employed disappeared during the fermentation, without disengaging any sensible quantity of nitrogen, he adopted the following experiment, which may serve as a type for all the rest.

In a solution of pure sugar-candy (we shall presently see that sugar and its analogues are the food necessary for yeast) we place, first, an ammoniacal salt—for instance, some ammonium tartrate—and then the mineral matter which enters into the composition of yeast, and add to this a slight, we may say imponderable, quantity of globules of fresh yeast. The globules sown under these conditions are developed, multiply, and excite fermentation of the sugar, while the mineral matter dissolves by slow degrees, and the ammonia disappears. In other words, the ammonia is transformed into the complex albuminoid matter which enters into the composition of the yeast, whilst the phosphates supply to the new globules their mineral principles: as for the carbon, it is evidently furnished by the sugar.

This, for example, is the composition of one of the liquids employed,—

10 grammes of pure sugar-candy,

Ashes of one gramme of yeast, obtained by means of a cupel-furnace,

0·1 gramme of ammonium dextro-tartrate,

Traces of washed beer-yeast, of about the size of a pin's head, in a fresh condition, damp, losing 80 per cent. of water at 100° C. (212° F.)

In such a mixture, the vessel being filled up to the neck, and well stopped, or furnished with a gas tube dipping into pure water, the fermentation began. After from twenty-four to thirty-six hours the liquor began to give evident signs of fermentation, by a disengagement of microscopic bubbles, which announced that the liquid was already saturated with carbon dioxide. On the following days the troubling of the liquor increased progressively, as well as the disengagement of gas, which was considerable enough for the froth to fill the neck of the flask. A deposit began, by degrees, to form at the bottom of the vessel. A drop of this deposit, examined under the microscope, shows a beautiful example of yeast, very much ramified, extremely young in appearance; that is to say, the globules are swollen, translucent, not granulated, and we distinguish among them, with surprising facility, each globule of the small quantity of yeast sown at the commencement of the experiment. These latter globules have a thick envelope, defined by a darker circle; their contents are yellowish; they are granular; but the manner in which they are sometimes surrounded by the newly formed globules shows very clearly that they have given rise to those among the latter which form the first links of the chaplets. If the observations are made in the earlier days of the formation, during the evening, and by gas-light, the old globules are distinguishable among the infinitely more numerous young ones, as you would distinguish a black ball in the midst of a great number of white ones.

This fundamental experiment proves, then, that yeast can bud, multiply, live and feed in a medium in which nitrogen is only represented by ammoniacal salts.

Further, M. Pasteur draws this conclusion from it that yeast causes with ammonia the synthesis of albuminoid substances. In order to test this conclusion, M. Pasteur weighed the ammonia which remained in the liquid some weeks afterwards, and found it less—slightly so, it is true (·0062 grammes). The weight of the newly formed yeast rose to ·043 grammes.

In his last memoir on the origin of muscular force, (Ann. Chim. Phys. [4], vol. 23, p. 5, 1871), Liebig energetically attacks the conclusions and results announced by Pasteur. He absolutely denies the formation of yeast, and its increase in weight, under the conditions of Pasteur's experiments, saying that he had never succeeded in realizing it; while, by substituting for the ammoniacal salts water in which yeast had been washed, the formation of fresh yeast was very evident. One of the arguments of Liebig, on which he relies most, is the absence of sulphur in the nutritive medium used by Pasteur; the albuminous substances contain some, and, therefore, yeast cannot elaborate proteinic matter under these conditions.

Let us remark, on this head, that sulphur occupies but little space in the complex molecule of albumin, and that nothing proves that this body is indispensable to the constitution of albumin.

In reply to this attack, M. Pasteur could only repeat, with forcible conviction, his former affirmations, and propose to his opponent to have the facts investigated by scientific referees. We will dwell no longer on this dispute, which was unfortunately terminated by the death of one of the most distinguished chemists of our age. We will only remark that fresh facts, studied with great attention, especially by M. Raulin, have, by analogy, given complete comfirmation to M. Pasteur's opinions concerning the nutrition of simple organisms in general, and yeast in particular. We shall allude to these facts later on.

M. Pasteur's experiment, as described by him in his memoir on alcoholic fermentation (Ann. de Chimie et de Phys., vol. 58 [3], p. 390), may be liable to this criticism—that, in proportion to the small quantities of yeast at first employed, the weight of ammonia that had disappeared, and that of the newly formed yeast, are very minute; and that they might be considered as falling within the limits of experimental errors, if one did not know the great ability of this eminent observer.

These very small initial quantities of yeast were employed to avoid the suspicion that the nourishment f the new cells had been effected at the expense of the soluble principles excreted from the old ones (the water in which yeast has been washed is, in fact, very effectual in giving activity to the multiplication of the cells of *Saccharomyces*). The experiments made by Duclaux in a different manner prove that these objections have no real foundation, and that the ammonia of the medium actually disappears.

This skilful chemist introduced into a certain volume of a solution of sugar, 2·501 grammes of yeast, containing 0·215 grammes of nitrogen; the liquid also contained 1 gramme of ammonium tartrate, corresponding to 0·152 grammes of ammonia. After the fermentation, 2·326 grammes of yeast, containing 0·148 grammes of nitrogen, were obtained from it. The liquid contained 0·055 grammes of ammonia, and 0·170 grammes of nitrogen in the form of organic compounds; which gives the following balance of nitrogen:—

	Nit. before fermentation.	Nit. after fermentation.
In the yeast	0·215	0·148
In the composition of ammonia .	0·152	0·045
Under the form of nitrogenous organic matter, dissolved in the liquid		0·120
Total . . .	0·367	0·363

The two amounts agree, within an error of 4 milligrammes (·06 grains Eng.). We see clearly that three-fourths of the ammonia has disappeared, and we find it in the yeast and the surrounding liquid, under the form of nitrogenous organic combinations. This experiment having since received the confirmation of every well-observed fact, we may admit, with confidence, that yeast can effect the synthesis of its proteinic materials, at the expense of sugar and ammonia.

M. Mayer has proved, as a complement of the experiments of M. Pasteur and M. Duclaux, that the ammonium tartrate may be replaced by other ammoniacal salts, as the nitrate oxalate, &c.,* without any disadvantage to the nutrition and to the disappearance of ammonia; and thus, far from being decomposed, in order to supply ammonia during its development, and during the fermentation, as Döbereiner had asserted, the yeast consumes the ammonia contained in the liquids which are fermenting.

Although a salt of ammonia may serve for the nutrition and development of ferment, it is proved, however, by generally observed facts, that it is not the especial nitrogenous aliment for this simple organism. If we substitute for the ammoniacal salt in Pasteur's experiment natural juices (as that of grapes, or of beetroot, or the water in which yeast has been washed), containing nitrogenous organic substances, the quantity of yeast formed and decomposed in the same time

* In all the observations made by Pasteur and the other experimentalists, an increase or decrease of the rate of development of the cells, their multiplication and their nutrition, was always accompanied by a variation in some direction in the energy with which the sugar (one of the essential nutritive elements) was resolved into alcohol and carbon dioxide. We will discuss presently what explanation may be given of the correlation between the two phenomena.

is much greater, and the decomposition of sugar is more active.

There exist, therefore, nitrogenous carburetted substances which are better suited to the nutrition of yeast than ammonia. What are they?

The natural juices which we have just mentioned, and more especially yeast-water which shows itself to be particularly active, contain different kinds of nitrogenous matter, and particularly albuminoid substances. Direct experiments alone can determine whether we are to attribute the active part played by these juices to proteinic principles, or to more simple compounds.

Pasteur found that the albumin of the white of egg was entirely unfit for the nourishment of the globules of yeast. Even at an earlier period M. Thénard and M. Colin had observed that albumin does not begin to excite alcoholic fermentation (or, which is the same thing, to produce the nutrition and development of the yeast globules) till the end of three weeks or a month, when left at the temperature of 30° C. (86° Fahr.), when it undergoes, under the influence of infusoria and mucidines, which are developed in it, a greater or less degree of decomposition. The serum of blood encourages the nutrition of the globules, without requiring to be itself previously decomposed; but it is not the serine which is active in this case; for, if we eleminate it by coagulation, the boiled and filtered liquid, when sugar and yeast are added, rapidly produces an energetic fermentation (Pasteur).

M. Mayer has made many experiments, with the view of throwing light on the question of the nutritive part played by albuminoid substances. He has found that the inactivity of the greater part of them (albumin, casein, &c.) arises especially from their not being diffusible through the organized membranes of the cells. We know, in fact, that most of these bodies belong to the class of colloid substances, not diffusible through porous membranes, and we can easily understand that, not being able to penetrate into the interior of the cell, but remaining imprisoned in the surrounding liquid, they will not be in a condition to give any powerful aid to the development, the nutrition, and the multiplication of the globules. The diffusible products formed by stomachic, intestinal, or artificial digestion, show themselves to be eminently suited to nourish the cell of the *Saccharomyces.*

It is the same with regard to diastase, and the different kinds of pepsin; but, as the nutritive activity of these ferments remain after they are cooked, it must by no means be attributed to their specific properties as soluble ferments (for these properties are destroyed by heat), but rather to products analogous to peptone, which always accompany these soluble ferments, and of which it is very difficult to get rid—syntonin, and, in a less degree, allantoin, urea guanine, and uric acid, increase the fermenting power of yeast; that is to say, give nourishment to the organism.

Other nitrogenous substances, which we ought also to consider as compound ammonias, have shown themselves to be but slightly, or not at all, active. Such are creatin, creatinine, caffeine, asparagin, leucine, hydroxylamine.

It is, then, very probable, according to these results,

that natural juices, the wort of beer, and water in which fresh yeast has been washed, owe their power of nourishing the cells of yeast, not to albuminoid principles, properly so called, which are indiffusible, but to allied nitrogenous compounds analogous to peptones, which have the property of passing by osmose through membranes.

Yeast thus affords us an evident and striking example of vegetable cells, which assimilate their nitrogen under the form of complex combinations, allied by their constitution to the higher forms of albuminoid substances. There is no proof that similar phenomena are not produced in plants of higher organization.

Agricultural and physiological experiments do not establish so clearly the assimilation of nitrogenous organic combinations as that of nitrates and ammoniacal salts; although the few observations made on this subject are rather favourable than otherwise to a positive solution of the question. But, if it were otherwise, it would not be logical and prudent to seek to draw a sharp distinction between the phenomena of nutrition of the *Saccharomyces*, and those of larger plants, and to say that the former may find nitrogenous nourishment in the organic nitrogenous combinations allied to albuminous substances, and that the latter do not.

Plants of complex organization are, in fact, formed by the union of cellular elements of various kinds, fulfilling different functions, and whose conditions of nutrition and development are not identical; among which it is probable that some are to be found susceptible of assimilating complex nitrogenous organic materials, that have been elaborated elsewhere at the expense of ammoniacal salts, or of nitrates.

Assimilation of Mineral Principles.—Vegetables, in general, always leave, after combustion in air or in oxygen, a fixed mineral residuum, whose weight varies, within certain limits, from one vegetable species to another, and also, especially, from one organ to another, as well as for the same organ, according to its age. Thus, the leaves of the pear-tree yielded to M. Violette, 7·118 grammes of ash per cent. of dried matter :—

The extremity of the twigs	bark	3·454
	wood	0·304
The middle part	bark	3·682
	wood	0·134
The lower part	bark	2·903
	wood	0·354
The trunk	bark	2·657
	wood	0·296
The roots	bark	1·127
	wood	0·234

In proportion as the plant grows older, the weight of the ash increases.

Do, then, mineral principles, which are found in all vegetables, from the highest point of the scale to the lowest—for we have already seen, by analysis, that yeast forms no exception—do, then, mineral principles play an important part in the biological phenomena of the nutrition and development of the plant, or do they only make their appearances as useless, but not injurious elements, inevitably introduced by the fluids from which the plant derives its nutritive principles ?

The remarkable constancy of the chemical composition of various kinds of ash, especially with respect to their constituent elements, as well as the most complete agricultural experiments, have proved in the most positive manner that the greater part of saline compounds, found by analysis, are necessary to vegetation. We have also learned to classify them in the order of their nutritive importance, with reference to the entire plant and its different constituent parts (leaves, stalks, seeds, grain, &c.).

Thus the phosphates preponderate remarkably in grain, and form by themselves the whole of the mineral mass found after incineration, as shown in the following results, published by Berthier :—

PHOSPHATES IN 100 PARTS OF ASH.

Nature of Phosphate.	White Wheat—Chartrain.	Rye.	Barley.	Oats.	Rice from Carmargue.	Maize.	French Beans—Soissons.	Green Peas.	Lentils.
Potassium phosphate	50·00	48·50	52·50	7·50	24·10	41·50	42·70	66·70	61·70
Calcium ,,	22·00	29·20	15·00	16·50	24·10	18·50	8·40	22·20	6·50
Magnesium ,,	28·00	—	25·00	20·00	24·10	38·00	14·30	6·60	19·60
Manganese ,,	—	18·30	—	—	—	—	—	—	—
Total	100·00	96·00	92·50	44·00	72·30	65·50	98·09	95·50	87·80

We also give, for reference, from the same author, the analyses of the ashes of various parts of plants.

STALKS, 100 PARTS OF ASH CONTAIN—

Composition of the Ashes.	Vine of Nemours.		Wheat Straw	Lucern.	Hay.
Potash	—	—	3·40	—	—
Potassium and sodium carbonates	16·40	—	—	14·44	12·20
Potassium chloride	2·20	0·78	2·50	1·90	3·64
,, sulphate	4·40	3·40	0·30	2·66	1·30
,, phosphate ..	—	—	—	—	—
,, silicate	—	4·00	—	—	—
Lime	—	—	15·70	—	—
Calcium carbonate	49·82	6·00	—	64·26	22·62
Magnesium ,, ...	3·85	—	—	6·07	6·39
Carbon dioxide	—	1·00	—	—	—
Iron oxide	—	—	2·60	—	—
Calcium phosphate	15·70	6·60	9·00	8·43	11·31
Magnesium ,, ...	—	—	—	—	—
Iron ,, ...	1·83	—	—	—	—
Manganese ,, ...	—	—	—	—	—
Phosphoric acid	—	—	1·20	—	—
Silica	5·80	78·22	73·90	2·24	39·80

BULBS AND ROOTS GAVE, FOR 100 PARTS ASH—

	Madder Root.	Jerusalem Artichoke.	Potatoes.	Onions.
Potassium and sodium carbonates	31·11	31·50	42·43	21·60
Potassium chloride	3·14	7·50	4·00	2·20
Sodium ,, ...	—	—	—	—
Potassium sulphate	3·93	6·00	2·80	4·00
,, phosphate	—	30·00	34·70	—
Calcium carbonate	35·01	—	2·80	12·00
Magnesium ,, ...	4·13	—	—	10·00
Calcium phosphate	9·71	16·50	6·87	38·00
Magnesium ,, ...	—	8·50	2·50	—
Iron ,, ...	5·09	—	1·70	—
Silica	7·88	—	2·50	—

In the leaves, the calcium carbonate and the silica preponderate, and form in themselves from 60 to 90 per cent. of the total weight of ash.

	Pine Leaves.	Vine Leaves.	Mulberry Leaves.
Calcium carbonate	68·74	51·00	53·00
Silica	6·43	10·20	27·70
Total	75·17	61·20	80·70

Let us compare with this the composition of the ash of beer-yeast (Mitscherlich):—

	Surface Yeast.	Sedimentary Yeast.
Phosphoric acid . . .	41·8	39·5
Potassa	39·8	28·5
Soda	—	—
Magnesium phosphate . .	16·8	22·6
Calcium " . .	2·3	9·7

If mineral salts really play an active part in vegetation, we may foresee, from these analyses,—

First. That their relative importance will vary with the respective weights of these different bodies, found in organized tissues; that, in consequence, the phosphates, potash, soda, magnesia, and the sulphates will occupy the first place.

Secondly. That according as the soil or the medium in which the plants grow is more especially manured by any one of these salts in particular, the development either of leaves, stalks, or seeds will be favoured. These results have all been verified by direct experiment, but as it does not enter into our plan to study agricultural chemistry with reference to mineral manures, and as our intention is merely to compare the nutrition of yeast, and of analogous organisms, with that of other plants of a higher order, we will return to the history of the *Saccharomyces*. We would remark that the composition of its ash, entirely composed of phosphates, approaches more nearly to that of seeds, to which it is also allied by the analogy of function and general chemical composition.

We owe to M. Pasteur the proof of the absolute necessity of mineral salts (phosphates of the alkalis, and the alkaline earths) for the development and nutrition of the yeast-cell. If, in his experiment, in which the ferment is sown, in an imponderable quantity, in a medium entirely composed of pure sugar-candy, ammonium tartrate, and ash of yeast, we omit the latter element, the fermentation and development of cells which ought to precede it (*i.e.*, the fermentation), do not take place. M. Pasteur went no farther in these researches; absorbed by the pursuit of a different aim, he did not endeavour to ascertain what were the most favourable mineral substances; he only used in his researches ash of fresh yeast as an inorganic element, rightly thinking that, at all events, he should find in it that which agrees best with the mineral nutrition of the fungus.

M. Mayer, following up this work, and seeking to ascertain by direct experiment which among the salts generally contained in varying proportions in vegetable ashes clearly favour the development of yeast, arrived (*loc. cit.*) at the following conclusions:—

1. Preparations of iron, employed in very small quantities, seem to have no influence; in larger proportions they are injurious.

2. Potassium phosphate shows a preponderating favourable influence. It may be employed in a liquid medium, in a high percentage, without its fertilizing influence being destroyed; while, for plants of a higher order, so great a concentration would become a serious cause of pathological disturbance. Potassium phosphate is not only favourable, but indispensable.

In fact, if from a medium formed of sugar-candy, ammonium nitrate, traces of yeast, and a mixture of acid potassium phosphate, magnesium sulphate, and tricalcic phosphate, a medium which ferments with considerable activity, we omit the acid potassium phosphate, fermentation and the development of yeast are not produced.

Potassium phosphate can by no means be replaced by sodium phosphate, which is inactive.

The absence of calcium phosphate from the medium causes much less injurious consequences than the omission of the before-mentioned salt.

The result is, that potassium and phosphoric acid are indispensable elements, whilst lime may be omitted without any great inconvenience, as we might have foreseen from the results of the analysis.

Magnesium, on the contrary, appeared in Mayer's experiments to be a very useful, if not an indispensable, element. It is immaterial whether this metal is supplied under the form of sulphate, or of ammoniaco-magnesian phosphate.

The combinations of sodium present no material effects, conformably with what has been already observed in plants of a higher order.

The sulphur, administered to yeast under the form of sulphates, or soluble sulphites, appears not to be assimilated. At least, the presence or absence of these two classes of salts seems to have no influence. Yet yeast contains sulphur in appreciable proportions, which we even find combined intimately in the products of its dis-assimilation (sulphuretted pseudo-leucine of Heintz). We are unable to say what the origin of this normal sulphur may be.

M. Raulin, in a remarkable investigation, has studied with particular care, and by an excellent method, the influence of the mineral components of the medium on the development of a cellular plant, the *Aspergillus niger*. As the results obtained may be interesting with respect to the question, rather as a general one than specially relating to yeast, which we are now considering, we will enter into some details on this point, more especially since the experimental method employed by M. Raulin may perhaps serve as a model for other researches of this kind.

First of all, an artificial medium is prepared, exclusively formed of definite chemical compounds suitable for the vegetation of a particular plant. In order to study the influence of various physical or chemical circumstances on the development of this plant, a vessel is filled with the artificial mixture and placed under the most favourable conditions for its vegetation. The seeds of the plant are sown in it, and they are allowed to grow during the necessary time; this trial, which is reproduced exactly in the same manner in each series of experiments, is the typical trial, with which all others are compared.

Another experiment is arranged in every respect like the first, with the exception of the single circumstance which it is proposed to study. The two crops, obtained at the same time, are dried and weighed separately, and the numerical ratio of the weight of these two results will be the measure of the influence of the condition to be examined.

The degree of perfection of the method depends upon three general conditions:

1. It is absolutely necessary to find, in the first place, an artificial medium suited to the development of the plant to be studied. M. Raulin found the ground quite prepared in this respect, thanks to the labours of M. Pasteur: the latter had observed that the mucidines (*Penicillium*) can be developed in a medium exclusively formed of definite artificial substances.

Water, sugar, ammoniacal salt (bitartrate), and ash of yeast. No portion whatever of the constituent parts of this medium can be omitted, without giving a complete check to the development.

2. The weight of the crop which the medium intended for the typical experiments can yield in a given time, with a constant weight of nutritive substances, ought, all other things being equal, to be as great as possible.

3. The typical experiments placed under the same conditions ought to yield crops whose numerical ratios differ but little from that assumed as unity: the ratio

which differs the most fixes the relative maximum error of the process.

At the beginning of his researches, M. Raulin, relying on the data of M. Pasteur, made use of a typical medium composed of—

Water	2,000
Sugar	70
Ammonium nitrate	3
Tartaric acid	2
Ammonium phosphate . . .	small quantities.
Potassium carbonate . . .	
Calcium „ . . .	
Magnesium „ . . .	

Aspergillus sowed at 20° C. (68° Fahr.).

With such a medium, the variation in weight of the crop, between one typical experiment and another, was so considerable, that it was not possible to ascertain the influence exercised by the omission of certain elements, which entered into the mixture only in small proportions. Thus, after forty-eight hours' vegetation, the weight of two typical crops were found equal to—

	Grammes.		Grammes.
No. 1 . . .	3·19.	No. 2 . . .	1·77.

By the omission of all the mineral elements he arrived at the result—

	Grammes.		Grammes.
No. 1 . . .	0·10.	No. 2 . . .	0·87.

The omission of potassium carbonate alone gave—

	Grammes.		Grammes.
No. 1 . . .	2·29.	No. 2 . . .	1·11.

The action of the whole of the mineral salts comes out strongly; that of the potassium carbonate is not perceptible, for the number 2·29 lies between 3·19 and 1·177, the numbers found for the typical experiments.

Besides, in these first experiments, M. Raulin ascertained that the development of the mucidines was fairly rapid in the first few days, and then grew indefinitely slower. While seeking to find the causes of this disturbance by tentatively modifying the conditions of the medium, especially by adding to it sulphur, zinc, iron, and silicon, in the form of salts; by modifying the proportions of the essential elements, raising the temperature to 35° C. (95° F.); and, finally, by employing vessels of considerable area and small depth, he succeeded in finding a typical medium, giving for the same length of time a result fifty times greater than that of the first experiments. Under these conditions, the ratio of the typical experiments, instead of varying from 1·0 to 1·8, acquired a remarkable constancy, and did not vary now more than $\frac{1}{24}$ of its value. It is evident that the favourable influence of any particular substance will then show itself in a much more defined manner.

The experiments, as far as *Aspergillus niger* is concerned, must now be conducted in the following manner:—

In the vessel intended for the typical experiment, the following chemical substances are brought together:—

Water	1,500
Sugar-candy	70
Tartaric acid	4
Ammonium nitrate	4
„ phosphate	0·60

Potassium carbonate	. . .	0·60
Magnesium ,,	. . .	0·40
Ammonium sulphate	. . .	0·25
Zinc ,,	. . .	0·07
Iron ,,	. . .	0·07
Potassium silicate	. . .	0·07

This mixture is left to itself for several hours, and then stirred with a porcelain spatula.

In order to sow the fungus, it is sufficient to pass over the whole surface the end of a camels'-hair brush with which spores have been collected from a very pure, and not too dry, vegetation of *Aspergillus.*

When we are not yet in possession of any *Aspergillus,* it is sufficient, in order to procure this plant in a pure state, to leave exposed to the air certain natural substances, such as the water of acidulated yeast, damp bread, or slices of lemon. The spores of *Aspergillus* which exist among the germs in the atmosphere may fall on these matters, and develop themselves there, mixed with other organisms.

When we see *Aspergillus* make its appearance, which is immediately distinguishable by its black fructification, it is again sown on an artificial liquid, and we at last obtain it free from mixture. The typical experiment being thus prepared, it is placed in a stove at 35° C. (95° F.), and constantly supplied with damp air. The spores develop, and at the end of twenty-four hours the filaments of the mycelium form a continuous whitish membrane on the surface of the liquid At the end of forty-eight hours this membrane has become very thick, and turns to a deep brown colour; after three days it has become quite black on the upper surface, which colour is due to the appearance of spores. The thick membrane is then removed by the fingers, squeezed, and then spread upon a plate to dry. New spores are sown upon the liquid, and after three days we obtain a second crop, weaker than the first.

The typical mixture, and that which is to serve for the experiment, and which differ from each other only in the single element, the influence of which we wish to ascertain, are placed together in the stove. This influence is measured by the ratio of the weights of the two first crops, or, still better, by the amount of the first and second crops obtained in six days.

Example.—The following trial mixtures were placed in the stove:—

No. 1. Typical medium.

No. 2. " " less the potass.

	No. 1. Grammes.	No. 2. Grammes.
First crop (after 3 days) . . .	14·4	0·80
Second crop (after 3 other days) .	10·0	0·12
Total	24·4	0·92

Ratio of the two first crops $\frac{0\cdot8}{14\cdot4}=\frac{1}{18}$; that of the weights of the whole crops $\frac{0\cdot92}{24\cdot4}=\frac{1}{26}$, numbers which prove plainly, in the most complete manner, the utility of the potass.

The results obtained by this remarkable method are the following:—

1. All the elements of the typical artificial medium concur simultaneously in the development of the plant, for if we omit each of them in turn, the weight of the crop undergoes a diminution, which is usually somewhat

considerable, and which cannot be attributed to experimental errors.

2. The mineral oxides of the artificial medium cannot be substituted for each other.

3. Nitric acid may be used instead of the ammoniacal salts as the nitrogenous aliment.

Finally, the following are the ratios found between the typical and the experimental trials :—

Omission of the oxygen	very great
" " " water	infinite
" " " sugar	65
" " tartaric acid	infinite
" " ammoniacal salt, or nitrate . .	153
" " phosphoric acid	182
" " magnesia	91
" " potash	25
" " sulphuric acid	24
" " zinc oxide	10
" " iron oxide	2·7
" " silica	1·4

The nutritive elements of the artificial medium are some indispensable, which are those found in large proportions in it, and others are useful, but apparently not indispensable : these only enter into the composition of the typical medium in very small proportions.

It is probable that some of these, such as sulphur, exist accidentally in very small quantities in the artificial media, to which they have not been added, and may thus set up a sluggish development. Thus, Mayer asserts that he has been unable by repeated crystallizations, and even by precipitating the liquor by barium chloride, to obtain sugar free from sulphur ; it is to the presence of this sulphur that he attributes the introduction of this element into the newly formed yeast.

It is evident that the results obtained by M. Raulin, and especially his method, may be applied to the search after better conditions for the maximum development of other kinds of vegetation, or simple organisms. M. Pasteur, who in his laboratory at the "Ecole normal" discovered new methods for the production of pure beer-yeast on a large scale, had to make preparatory trials analogous to those of M. Raulin. His researches also establish an important fact, that an artificial medium, suitably prepared, may be as favourable to the development of vegetation, and even more favourable, than the most fertile natural media. We may thence draw conclusions of great importance as to the cultivation of larger plants, and may suppose that chemical manures, suitably chosen, may be substituted for natural ones in agriculture, with great advantage.

This is what several men of science, who make a study of agriculture, have already attempted; the great point is to determine carefully the useful composition of these manures; unfortunately, we must admit, experiments on larger plants are not so simple and so easily managed as those on mucidines.

Sugar.—Pasteur and Raulin have demonstrated the preponderating influence of, and the *necessary* part played by, sugar or analogous bodies in the vegetation of *Aspergillus* and of mucidines. This influence is as powerful in the development of the yeast of beer. Without sugar, without hydrocarbonate substances, yeast can neither reproduce nor be nourished. An important difference is thus, at first sight, established between simple organisms, such as ferments, mildews, &c., and the larger plants, which derive the organic elements of their constitution from the simplest compounds of carbon, such as carbon dioxide.

This distinction, however, loses its force after a more complete examination.

If the larger plants derive nourishment at the expense of carbon dioxide, it is because, in their leaves and other green parts, there are found organs suited to the utilization of the active force of the luminous rays sent by the sun or other source of light. The carbon is set free directly, and the oxygen is disengaged. It appears very probable, that at the moment when the carbon is separated from the oxygen, under a special condition as yet unknown, and very different from that of black amorphous carbon, or of the diamond or graphite (forms under which we know this element), that it unites with the elements of water to form a hydrate of carbon (starch, sugar ?), or at least a body which can be converted into these principles by ulterior transformations.

If we were able to effect the decomposition of carbon dioxide under the influence of light outside the animal economy, I have no doubt but that (if the experiment were made in the presence of water) there would be found a hydrocarbon compound. I have even been able to give a slight experimental confirmation of this theoretical opinion.

If we treat, in the cold, coarsely powdered white cast iron (which is known to contain an iron carburet), with a solution of cupric sulphate, the iron of the white cast iron is entirely dissolved, without disengagement of any carbon or other gas ; after having washed it, we may eliminate the deposited copper by placing it in contact with a solution of iron perchloride. The copper is rapidly dissolved; there remains a pulverulent black mass, which, after dessication at 80° C. (176° Fahr.), in a vacuum, resembles carbon. But this carbon contains water in combination, which is suddenly disengaged when it is heated to about 250° C. (480° Fahr.); it is easily dissolved in nitric acid, becoming oxidated, yielding yellow or orange-yellow substances containing nitrogen. This residuum, when analyzed, gives a quantity of water, which is in a tolerably constant proportion to that of carbon.

It therefore represents a true and definite carbon hydrate. It is evident that the condition of the carbon in the cast iron must be very different from that of the carbon of carbon dioxide, and that the hydrates which arise from the separation of these forms of carbon may differ greatly. Nevertheless, the experiment which I have just described gives material support to the idea which physiologists entertain of the successive chemical metamorphoses of the carbon compounds in plants.

When the carbon hydrate is once formed in the leaf, it is carried into the other parts of the plant, to serve there as nutrition, for the development of cells containing no chlorophyll, and whose biological functions closely resemble those of cellular organisms. That which takes place during the germination of seeds, up to the moment when the new plant becomes provided with aerial leaves which have become green under the influence of light and air, and begins to utilize carbon dioxide, leaves no doubt as to the scientific value of this interpretation.

We see here the newly formed cells successively developed, and superposed so as to form radicles, stalk, cotyledons, and leaves, and the germ procures the necessary materials for its development from the organic principles which the seed has accumulated, and among which hydrocarbons are always predominant.

We must, therefore, admit that the phenomena of nutrition of the larger plants do not seem to differ much, when examined in detail, from those of the more simple ones.

The former are provided with special organs which enable them to elaborate for themselves the hydrocarbon substances which they require for the development of the rest of their organism. The inferior cellular plants, and in fact generally all those which are unprovided with cells containing chlorophyll, are necessarily parasites, which must borrow their hydrocarbonate nourishment, directly or indirectly, from plants furnished with these cells.

Besides these general considerations, founded on the phenomena of nutrition observed in plants, the experiments of M. Pasteur establish with certainty, that in all alcoholic fermentation a part of the sugar is fixed in the yeast, in the state of cellulose or some analogous body. In fact, since infinitely small quantities of yeast, sown in a medium entirely formed of pure sugar-candy, of ammonium tartrate or nitrate (Mayer), and ash of yeast, develop and give rise to very ponderable proportions of yeast, considerably greater than the original quantities, it cannot be doubted that the hydrocarbon principles of this new vegetation (cellulose, &c.), are furnished by the elements of the sugar.

The following experiments lead to the same result:—

M. Pasteur submitted to fermentation 100 grammes of sugar, about 750 cubic centimètres of water, 2·626 of yeast (weight of the dried matter).

After the fermentation, which lasted twenty days, he collected 2·965 grammes of yeast (dried matter).

He likewise boiled for six or eight hours a determined weight of fermented yeast, and also of the same yeast before fermentation, with sulphuric acid, diluted with twenty times its weight of water (fermented yeast 1·707 grammes, and unfermented yeast 1·73 grammes, dried at a temperature of 100° C. (212° F.).

The insoluble residues were weighed on filters, the weight of which had been estimated; they were then washed, dried at 100° C. (212° F.), and weighed. The filtered liquids were neutralized with barium carbonate; the quantity of sugar formed by the action of the sulphuric acid on the cellulose was ascertained, either by means of Fehling's liquid, or by fermentation.

He thus found, by calculating the results obtained for the two weights, 2·626 grammes, and 2·965 grammes of yeast employed, and yeast obtained—

1. That the 2·626 grammes of crude yeast employed gave an insoluble nitrogenous residuum equal to 0·391 (14·8 per cent.), and 0·532 of fermentable sugar.

2. That the 2·965 grammes of yeast found after fermentation leave a nitrogenized residuum of 0·634 grammes (about 21·4 per cent.), and 0·918 grammes of fermentable sugar.

There was, therefore, fixed in the fermentation of 100 grammes of sugar with 2·626 grammes of yeast, 0·4 grammes of hydrocarbon matter, transformable, by dilute sulphuric acid, into fermentable sugar; there was also a sensible augmentation of nitrogenous matter, insoluble in dilute sulphuric acid.

On the other hand, in order to verify, by a second experiment, the value of these conclusions, M. Pasteur made use of the process of separating cellulose from the albuminoid substances indicated by Payen and Schlossberger. This process consists, as is well known, in treating yeast with dilute solutions of potass.

In three careful experiments, M. Pasteur found a residuum, insoluble in potass, formed of cellulose, transformable into sugar by being boiled with dilute sulphuric acid of 17·77, 19·29, and 19·21 per cent. of the dry yeast experimented on.

But the 0·532 grammes of sugar, produced without the intervention of potass, by 2·626 grammes of the same yeast, correspond to 20 per cent. of yeast. It is, therefore, proved that boiling in sulphuric acid had removed all the cellulose.

Let us also notice that the 2·965 grammes of yeast found after fermentation, giving 0·918 grammes of sugar, ought to contain 31·9 per cent. of cellulose, a quantity 11 per cent. greater than there was before fermentation. This considerable augmentation of the weight of the cellulose in the yeast, while it exercised its action on the sugar, is a point worthy of remark, since it proves that, in accomplishing one of its principal functions, yeast undergoes very marked evolutions in its composition.

The following experiment of M. Pasteur's proves, besides, that during fermentation the yeast itself forms its fatty matter by the help of the elements of sugar.

Let us first call to mind that Payen's analyses show 2 per cent. of fatty matter in the yeast, and that the lees of wine also contain fatty matter. It had been thought that this fatty matter was furnished by the fermentable medium. Pasteur mixed sweetened water (prepared with pure sugar-candy) with the watery extract of limpid yeast, heated several times with alcohol and ether. He sowed in it an imponderable quantity of fresh globules. These multiplied, and caused the sugar to ferment. He succeeded thus in preparing some grammes of yeast from substances completely without fatty matter. But this newly formed yeast does not contain less than from 1 to 2 per cent. of fatty saponifiable matter, yielding crystallized fatty acids. The same fact is observed with yeast which has been grown in a medium composed of sugar, water, ammonia, and phosphate. It is, therefore, from the elements of the sugar that the fatty matter is obtained.

These facts confirm the views of M. Dumas as to the possible formation of fatty matter from sugar.

Water.—Water is, we need hardly say, quite as indispensable for yeast, and the elementary organisms, as for higher forms of life.

According to Wiesner, the cell of yeast manifests its activity, develops, and is nourished within the limits of hydratation comprised between 40 and 80 per cent. of water. Yeast dried with precaution may regain its power when moistened afresh. It may be understood from this why a solution of sugar, the concentration of which exceeds 35 per cent., is not changed by ferment: such a solution takes from the cells, by osmose, a sufficient quantity of water to lower their hydratation below 40 per cent.

Wiesner's researches have also shown that there are two states of concentration in which the phenomena of the fermentation and nutrition of yeast attain their maximum value. One of these maxima corresponds to a solution of from 2 to 4 per cent. of sugar; the other to a solution of from 20 to 25 per cent. These facts require confirmation; at all events, there is at present no conclusion to be drawn from them.

Oxygen.—The cells of the *Saccharomyces cerevisiæ*, introduced into a liquid medium containing oxygen in solution (pure water, a saccharine solution, with or without nutritive mineral and nitrogenous elements), absorb oxygen with great rapidity, and develop a corresponding quantity of carbon dioxide. This fact, which constitutes true respiration, comparable to that of animals, has been brought to light by M. Pasteur. An excellent method of obtaining water completely deoxygenized, much more efficaciously than by boiling, consists in diffusing through the water one or two grammes (from 15·4 to 30·8 grains) per litre (1·76 pints English) of fresh yeast in the form of paste, and leaving the liquid undisturbed for from one to two hours at a temperature of 25° to 30° C. (77° to 86° F.). Powdered zinc shaken up with water, containing air in solution, gives the same results.

I determined, by the help of M. Quinquand, the weight of oxygen absorbed by the unit of weight of yeast, in the unit of time, when this organism is placed in water containing air in solution, without any mixture of nutritive materials. These measurements were taken by an oxymetrical process which I have invented, with the assistance of M. C. Risler, one of my pupils. As this method seems to me likely to be serviceable in researches of this kind, and in the study of biological phenomena, I think I ought to give the description of it here, even at the risk of introducing a foreign element into the examination of the facts which now occupy our attention.

Process for the Volumetric Estimation of Dissolved Oxygen.—The process of measurement of the quantity of oxygen dissolved in water, by means of a standard liquid, which was proposed by M. Gérardin and myself, (Comp. Rend., vol. 75, p. 879), and which I have since improved, with the assistance of M. C. Risler, depends essentially on the energetic reducing properties of sodium hyposulphite. This salt* is obtained with the greatest facility by the action of a solution of sodium bisulphite on zinc, either in plates, shavings, or powder. Its formation is much more rapid when the zinc employed is finely divided, and the points of contact between the metal and the solution of bisulphite are more numerous. Thus, with powdered zinc employed in sufficient quantity, and a very concentrated solution of bisulphite (marking 35° Beauné, and requiring from 5 to 7 per cent. of its weight of powdered zinc), an agitation of from three to five minutes is sufficient to complete the reaction. This takes place, with elevation of temperature, according to the following equation:—

$$\underset{\text{Sodium bisulphite.}}{3\,(SO.\ NaO.\ HO)} + ZN^2 = \underset{\text{Sodium hyposulphite.}}{S.\ NaO.\ HO} + \underset{\text{Sodium sulphite.}}{SO\,(NaO)^2} + \underset{\text{Zinc sulphite.}}{SO\,(Zn\ O)^2} + \underset{\text{Water.}}{H^2\,O.}$$

* P. Schutzenberger on a new acid of sulphur, Ann. de Chim. et de Phys., vol. 20, p. 251 (4).

If the bisulphite used in the experiment is concentrated, there are deposited, a short time after the cooling of the liquid, crystals of the double zinc and sodium sulphite, while the hyposulphite formed remains in solution, still mixed with the sulphites. The impure solution of hyposulphite (a mixture of hyposulphite and of sulphite of soda and zinc) may be employed as it is for the estimation; but it will only keep for any length of time when protected from the air, and in a very dilute state.

By adding to this liquid a suitable quantity of milk of lime, we precipitate the zinc oxide, and by filtration we obtain a solution very slightly alkaline, endowed, like the former, with very decided reducing power; possessing the property of keeping for a longer time, when not exposed to the air, especially in a state of great dilution, the form under which it is always used in the quantitative estimation of oxygen. Without going at farther length into the properties of sodium hyposulphite (*see* the memoir in the Ann. de Chim. et de Phys., before cited), I ought to particularize those which are especially utilized in this process.

1. The sodium hyposulphite, not saturated by lime, absorbs oxygen with great rapidity, whether in the form of gas, or in solution; its action is, in this respect, similar to that of sodium pyrogallate.

By fixing the oxygen the hyposulphite becomes acid, and is converted unto sodium bisulphite.

$$\underset{\text{Hyposulphite.}}{S. (Na. O) (H\ O)} + O = \underset{\text{Bisulphite.}}{S\ O. (Na. O) (H\ O).}$$

When saturated by lime, it still acts in the same manner on the gaseous oxygen, but more slowly; while it absorbs dissolved oxygen, instantaneously, and removes it from the oxygenated liquid with which it is mixed.

2. The hyposulphite, poured into a solution of ammonio-cupric sulphate, reduces the cupric oxide to the state of cuprous oxide, destroying the colour of the liquid, and then it reduces the cuprous oxide in its turn, precipitating metallic copper; the reduction is made at two intervals of time, and we are able, by employing more or less hyposulphite, to stop the process at the first stage, shown by the decolouration of the liquid.

3. The hyposulphite, whether acid or neutralized, instantly destroys the colour, by reduction, of the solution of Coupier's blue (aniline blue), and of sodium sulphindigotate (indigo carmine). These bleached solutions resume their blue tint when exposed to the air.

4. If to water containing oxygen in solution, and coloured blue by aniline blue or indigo carmine, we add, little by little, a dilute solution of hyposulphite, either saturated or not with lime, the reducing agent acts at first upon the dissolved oxygen, and does not destroy the colour until it has absorbed it.

Thus, by taking two equal volumes of water tinted blue by either of these colouring matters, saturating one with oxygen by agitation with air, and depriving the other of its oxygen by sufficiently prolonged boiling, we shall find that the latter loses it colour after the addition of a few drops of the hyposulphite, while the second requires, in order to effect this result, a much greater quantity of the reducing solution, and one in proportion to the quantity of dissolved oxygen.

Here, however, a very remarkable peculiarity presents itself, one which we ought specially to point out, because our ignorance of it would entail grave errors in the analysis. If we have some aerated water, and a suitably dilute solution of the hyposulphite, either saturated or not with milk of lime, we may previously determine the oxymetric value of the hyposulphite, that is to say, the volume of oxygen which is required to saturate the unit of volume of the solution; it is only necessary for this purpose to prepare a solution of ammonio-cuprous sulphate, containing 4·46 grammes (68·826 grains) of pure crystallized cupric sulphate to the litre (1·76 Eng. pints). Such a solution having been brought exactly to the bleached state, without precipitation of metallic copper, that is to say, being brought back to the state of solution of ammonio-cuprous-oxide, will have yielded to the reducing liquid half of the oxygen corresponding with the cupric oxide which it contains, about 1 cub. centimètre of oxygen (·061 cub. in.) for each 10 cub. cent. (·61 cub. in.) of the solution.

It is sufficient, therefore, to determine with precision the volume of the hyposulphite necessary to decolourize completely 10 cubic centimètres of the cupric liquor without precipitation of metallic copper; this volume will correspond to one cubic centimètre of oxygen.

This being determined, let us colour with a little indigo carmine or aniline blue (just sufficiently to render the tint perceptible) a certain quantity (for instance, one or one half litre) of our aerated water, and let us pour in the hyposulphite with a burette; the moment will come when the last drop will effect the decolouration of the liquid, which will rapidly pass from blue to yellow. In this state, the clear yellow solution is a *very delicate* test of free oxygen ; it requires only the slightest bubble of air, of the size of a pin's head to produce very evident blue streaks. We are therefore induced to admit that the dissolved oxygen has been completely utilized by the reducing liquid. This is not, however, correct. If we calculate, according to the quantity of hyposulphite fixed by the cupric solution, and the volume of the reducing liquid employed to change the colour of the blue liquid, the quantity of oxygen contained in a litre of water, we find, as nearly as possible, half the oxygen really contained in this water, and the mercurial pump or boiling could disengage from it. This remarkable result has been determined by a great number of experiments. What, then, has become of the other half ?

M. Risler and I thought, at first, that the products of oxidation of the sodium hyposulphite were not the same when the oxidation took place under the influence of free oxygen, as under that of the ammonio-cupric oxide ; however, after having ascertained that in both cases sulphite was formed, and nothing but sulphite, we were obliged to abandon this interpretation ; we could only believe that the diluted hyposulphite, acting, in the cold, on the dissolved oxygen, divides it into two equal parts, one of which is fixed in the reducing liquid, and the other unites with the water, forming oxygenated water or some analogous compound. This second half of the oxygen, which has been, as it were, rendered latent, acts no longer either on the hyposulphite or on the indigo (discoloured carmine). When I say that it no longer acts, I mean under the conditions of the experiment, which may be considered to be instantaneous, and at a low temperature.

In fact, if we keep the bleached liquid—provided that we have not used too little indigo (carmine)—for some time from access of air, especially if we raise its temperature to 50° or 60° C. (122° to 140° F.), we see it become blue again instantaneously, and throughout the whole mass. The experiment may be made in a vessel filled with an atmosphere of pure hydrogen to which is fixed the extremity of a Mohr's burette, containing the hyposulphite. If we now add a fresh quantity of the reducing fluid until the second decolouration, the same effect will be produced again, and until we have introduced a volume of hyposulphite nearly equal to that employed in order to attain the first term of decolouration.

These experiments are delicate. To make them succeed they must be completely protected from the access of atmospheric oxygen; hyposulphite which has been neutralized by lime, should be employed, and there should be a sufficient quantity of indigo. (For further details on this subject, see the Bulletin de la Société Chim. de Paris, vol. 20, p. 145, 1873.) They prove that the first action of the acid hydrosulphite (which may be considered instantaneous) on the aerated water coloured by indigo, only removes half the oxygen. The other half acts much more slowly on the reduced indigo, and, by its *intervention*, on the hyposulphite in excess. This action does not manifest itself at all if the solution is even slightly acid. In this case, the latent oxygen may remain almost indefinitely in the presence of a great excess of hyposulphite, or of the reduced solution of indigo carmine, without being fixed by it.

By employing water coloured blue, to which has been added a little oxygenated water ($H_2 O_2$), we produce with the hyposulphite alternate decolourations, followed by spontaneous recolourations, of the whole mass, which resemble, so as to be indistinguishable from them, those which take place in the experiments already described. This similarity, added to the want of any other plausible explanation, makes me think that the latent oxygen is really found in the liquid under the form of oxygenated water.

If we operate on a liquor rather acid than neutral, or on a neutral liquor employing only hyposulphite not saturated with lime, which becomes acid as it oxidizes; finally, by taking note of the previous observation in our calculation, that is, multiplying the quantity of oxygen found by 2, we arrive at very close and satisfactory results.

I will first describe a rough method, susceptible of being used anywhere, on the banks of a river, or in the country, but which can furnish only approximate indications, by giving the amount of oxygen within a quarter of a cubic centimètre per litre (·015 cub. in. per 1·76 pint).

Some acid hyposulphite may be prepared, instantaneously, by agitating with zinc powder a diluted solution of sodium bisulphite, prepared with supersaturating sodium carbonate by a current of sulphurous acid (the bisulphite at 35° Beaumé is a commercial product, and may be used). This bisulphite at 35° Beaumé is previously diluted with four times its weight of water and for 100 grammes of the diluted solution we employ 2 grammes of zinc grey (powdered zinc). The mixture and the agitation are to be made in a vessel nearly filled with the liquid. After five minutes, the solution must be filtered, and suitably diluted with water, so that in a preliminary trial, one litre of water agitated with air (saturated with oxygen under the pressure of $\frac{1}{5}$ of an atmosphere at the ordinary temperature) and tinted blue by some drops of a solution of aniline blue, or indigo carmine, may be decoloured by about 25 or 35 cubic centimètres (1·525 to 2·135 cubic in.) of the solution of hyposulphite.

The analysis requires nothing but a vessel with a large mouth, (a wide-mouthed bottle), holding about 1½ litre, a stirrer which will allow us to mix together the different layers of liquid without disturbing the surface too much, one of Mohr's burettes, furnished with a narrow tube at one end, fixed to the india-rubber tube of the pinch-cock, and arranged so as to be held mid-way in the water; also a glass bottle or jar holding a little more than two litres, graduated so as to indicate 1 litre. A litre (about 1¾ pints) of the water to be tested, is introduced into the wide-mouthed bottle, tinted with aniline blue or indigo carmine; then, the burette being filled with hyposulphite, and its lower end previously filled with the liquid plunging midway into the water in the wide-mouthed bottle, we allow the reducing liquid to flow in slowly, agitating the contents with the stirrer up and down, so as not to disturb the surface too much; the experiment is stopped at the moment that the decolouration takes place, and the volume employed is read off.

Immediately after, we proceed to the estimation of the hyposulphite exactly in the same manner, by employing 1 litre of the same kind of water which served for the first experiment, but after having previously agitated it for some minutes with air, in the large bottle, and taken its temperature. Under these conditions, whether the original water be above or below the limit of saturation for oxygen; we always succeed quickly in having water saturated with oxygen at the pressure of $\frac{1}{5}$ of an atmosphere (the pressure of oxygen in the air), and at the temperature which has been read off. Tables of solubility, notably those of Bunsen (Méthodes Gazométriques, Traduction Française de T. Schneider) give the amount of oxygen.

Thus, in two experiments made under conditions identically similar, we have the volume of the reducing fluid required by the water whose oxygen is unknown, and that required by the water whose oxygen is known. A simple proportion will give the value of x in the problem. This process of estimating the hyposulphite, on account of its simplicity and certainty, is preferable to the employment of an ammoniacal solution of copper, which M. Gérardin and I had proposed; it was suggested by M. Raulin, assistant director of the laboratory of M. Pasteur. As the operation is performed with contact of air, it is necessary to make the measurements as quickly as possible, and to operate on a large quantity of water (a litre), in order to counteract as far as possible the influence of the oxygen of the air. Besides, the method of analysis described above has the effect of neutralizing almost entirely this cause of error; the two operations being made under the same conditions, the error can only proceed from a slight difference between the conditions of the two experiments, such as their duration, or the greater or less agitation of the water.

I have succeeded, by the assistance of M. Risler, in applying a similar method of quantitative analysis to much smaller quantities of water, or oxygenated liquid; and by modifying the process of the operation, I have been able to ascertain by the reducing liquid, not the *half*, but the whole of the dissolved oxygen; this method is much preferable and more certain. In order to attain this double result it is only requisite; 1st. To make the analysis in a liquid completely protected from access of the oxygen in the air, by an atmosphere of pure hydrogen: 2nd. To introduce the aerated water that is to be tested, (a known volume, from 40 to 100 cubic centimètres (2·44 to 6·1 cub. in.) into a tepid, 40° or 50° C. (104° to 122° F.) neutral, or very slightly alkaline, but never acid, medium, formed by a solution of indigo carmine, just decoloured by hyposulphite which has been previously neutralized or rendered slightly alkaline by milk of lime. This yellow medium turns blue under the influence of dissolved oxygen; a quantity of blue indigo is re-formed, proportional to the amount of oxygen dissolved. If the preceding conditions of temperature and neutralization have been carefully observed, all the dissolved oxygen is utilized in oxygenating the reduced indigo, and there only remains to be estimated, by means of the hyposulphite, the volume of this reducing agent necessary to decolour the blue liquid. The same experiment is repeated immediately after, with the same volume of water, agitated at a known temperature, and calculation will give, as before, the volume of oxygen sought. If the liquid is acid, or becomes so in the process of analysis, the conditions leading to the formation of oxygenated water are at once present, and the results are deceptive, and always too low, approaching more or less nearly to the half of the whole amount of oxygen.

If, then, we operate on acid liquids, we ought to add previously to the yellow indigotic liquid a sufficient quantity of a diluted solution of ammonia to correct this disturbing condition. The principles of the experiment being known, I will enter into some details concerning the apparatus, and the preparation of the reagents.

1. *Sodium Hyposulphite.*—We prepare the acid hyposulphite in the manner described above. This is neutralized by milk of lime. For 100 grammes of concentrated bisulphite at 35° Beaumé, employed for the preparation of the acid hyposulphite, we shall make use, for this purpose, of 35 grammes of milk of lime, prepared from 200 grammes of lime previously slaked, to 1 litre of water. The saturation is effected with the acid hyposulphite, diluted, as I have before said, with four times its weight of water, but the quantity of milk of lime is calculated according to the weight of concentrated bisulphite that is employed. It is shaken up, left to settle, decanted, and then filtered; the liquid is kept in full and well-stoppered bottles.

When used, it is necessary to dilute this liquid with distilled water, until 50 cubic centimètres (3·05 cub. in.), of water saturated with oxygen require from 4 to 5 cubic centimètres (·244 to ·305 in.), of the reducing agent.

The solution thus prepared may be kept almost indefinitely at the required strength, if we take the following precautions. It is to be placed in a bottle containing about 1 litre (1¾ pints) filled to the neck, closed by a good india-rubber cork, perforated with two holes; in one of them is fixed a tube bent at right angles, whose extremity dips to the bottom of the bottle, and whose other end has an india-rubber tube, furnished with a Mohr's pinch cock; in the free end of the india-rubber tube, is fixed a piece of glass tube; in the second hole is fixed another tube bent at right angles, but which only goes about 1 or 2 centimètres (·39 or ·78 in.) past the cork. This tube is kept in permanent communication by means of a sufficiently long india-rubber tube, with a gas-jet kept turned on. It is proper to interpose between the gas-burner and the bottle a glass vessel like those employed by chemists for drying gases; this vessel is filled with pumice stone, saturated with a concentrated solution of sodium pyrogallate. By this means, we get rid of the oxygen which common gas always contains (from 1 to 2 per cent.) The burettes are filled with hyposulphite by aspiration, from below upwards. For this purpose, the india-rubber pinch-cock tube, which, in order to serve another purpose, ought to be sufficiently long, is placed in communication with the tube which goes to the bottom of the hyposulphite, and we draw in the liquid by the mouth, by means of an india-rubber tube, fixed by a cork and a bent glass tube to the upper part of the Mohr's burette. By this means, we avoid the agitation of the liquid in contact with the air. With these precautions, the strength of the test does not alter ; it is, however, prudent not to trust to this, but to ascertain it at every fresh experiment, which is very easily done.

Indigo.—We prepare beforehand ten litres of the solution of indigo carmine, by dissolving in this quantity nearly 200 grammes (6·42 oz. troy), of indigo carmine in the form of paste (sodium sulphindigotate). The liquid should be kept in blue or black glass bottles, sheltered from the light.

The estimation is made in a three-necked bottle, of 1 or 1½ litres in capacity (Fig. 19). One of the three lateral necks receives, through an india-rubber cork, the tube, Hy., bent at right angles for the admission of hydrogen. It is well that this tube should slide in its cork without too much friction, in order that it may be raised or depressed at will, without allowing air to enter.

The other lateral neck has an india-rubber cork, pierced with two holes. In one of these is fixed the end of a small funnel with a glass stop-cock, E, a bromide funnel ; this end should be sufficiently long to reach the bottom of the bottle.

In the other hole is fixed the tube which carries off the hydrogen ; this tube, being twice bent, has its free extremity plunged into the mouth of a test-tube, T, which is fixed there by means of a cork, also pierced with two holes ; the tube contains water, and the hydrogen in excess, after having passed through this water, escapes by a tube bent at right angles, fixed in the second hole of the cork.

The middle neck of the bottle holds as a fixture a stopper of cork or india-rubber, hermetically cemented in, and pierced with two orifices, in which are also permanently fixed the two pointed extremities of two Mohr burettes, H I, held above the bottle by a special

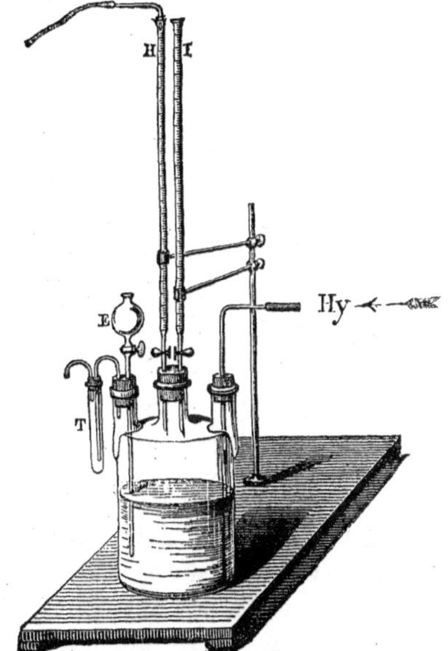

FIG. 19.—Apparatus for the measurement of oxygen dissolved in water.

support, one by the side of the other. The india-rubber pinch-cock tubes at the lower end of these two burettes ought to be long enough to allow the bottle to be moved without the burettes being shaken;

they are fixed to the burettes, and ought, on the contrary, to be able to be detached, at will, from the thin glass ends which pass through the middle cork of the bottle. It is well to have at hand a third burette, either independent, or fixed to the same support by the side of the others. It is intended as a reserve for the hyposulphite of the principal burette, H. One of the first two burettes, I, contains indigo carmine, the two others are filled with sodium hyposulphite.

This having been arranged, when we wish to measure the oxygen dissolved in any particular specimen of water, we introduce into the large bottle—

(1) About 50 cubic centimètres (3·05 cub. in.) of solution of indigo.

(2) About 250 cubic centimètres (15·25 cub. in.) of ordinary warm water, at 40° to 50° C. (104° to 122° F.). We adapt to the fixed sockets the burette, I, with the indigo, and the *independent* burette of hyposulphite; we charge the two sockets with the contents of the burettes, and allow the hydrogen to pass rather quickly. (The apparatus producing the hydrogen may be of any kind, provided that we are able to regulate the disengagement of gas by means of a single stop-cock; the hydrogen is purified by water, and passes through a column of caustic potash in plates.) When we suppose that almost the whole of the air has been swept away by the current of hydrogen, we allow the hyposulphite to flow in slowly, holding the bottle in the hand, and giving the liquid a gyratory motion sufficient to mix it, until it has only a slight greenish tint, or reddish if it is alkaline; then, without interrupting the current of hydrogen gas, we detach the independent burette, and fix on the hypo-

sulphite burette, H, intended for the analysis. In doing this, the test liquid contained in the socket escapes, and often completes the decolouration of the water by turning it yellow. We charge the socket again by slightly opening the pinch-cock, then by adding first a little carmine, then a little hyposulphite, we bring the liquor to a bright yellow colour, which verges towards green or red by the addition of a single drop of carmine. We notice that the yellow liquid does not turn blue at the surface, which proves that the atmosphere of the bottle is thoroughly deprived of oxygen. Then all is ready for the analysis if we have taken care, at first, to fill the lower end of the funnel with the same water as that whose contained oxygen we wish to measure. This estimation is made by proceeding in the following manner, and in the order indicated:—

(1) Slacken the current of hydrogen, without stopping it; (2) Raise the supply tube, so that the gas should not bubble up; (3) Read off the graduation of the hyposulphite burette, at which the surface of the liquid stands; (4) Introduce into the funnel 50 or 100 cubic centimètres of the water to be tested, and allow this water to run at once into the bottle, keeping the lower end full from the stop-cock; shake the bottle; the standard yellow liquid grows blue if the water is aerated, and does not change its colour if it contains no oxygen. We now need only allow the hyposulphite to run in, drop by drop, shaking it, and noticing the exact limit of decolouration which takes place, true to a single drop ($\frac{1}{20}$ of a cubic centimètre) if the liquid is without colour.

Immediately after having noted the point at which this takes place, without changing or disturbing anything, fill the body of the funnel with water saturated with oxygen, at a pressure of $\frac{1}{5}$ of an atmosphere, and at a known temperature; correct with some drops of hyposulphite the slightly blue tint developed by this operation, and proceed to estimate the value with 50 or 100 cubic centimètres of saturated water. A rule of three sum gives the quantity of oxygen in the first portion of water.

Instead of testing the hyposulphite with saturated water at a known temperature, we may estimate it by allowing a known volume (25 cubic centimètres) of solution of indigo carmine to flow into the bottle after the first trial, and ascertaining the volume of the reducing agent necessary to bring back the decolouration.

On the other hand, we have determined, once for all, the volumetric ratio between the indigo employed, and a solution of ammonio-cupric sulphate, containing 4·46 grammes of pure crystallized cupric sulphate to the litre; this amounts to 10 cubic centimètres = 1 cubic centimètre of oxygen at the moment of complete decolouration. It is sufficient for this purpose to ascertain the volumes of the same solution of hyposulphite necessary to decolour equal volumes of indigo carmine and of the cupric liquid, which allows us to calculate the ratios of volumetric equivalence of these two liquids, and consequently the volume of oxygen corresponding to 1 cubic centimètre of the indigo solution.

This experiment ought to be made with great care, for on its accuracy will depend that of the absolute values of all the subsequent analyses.

We determine the ratio between the hyposulphite and the indigo under the same conditions as those of the preceding estimation, since these are the conditions to which we shall always come back in future experiments.

The ratio between the same hyposulphite and the cupric solution is determined by operating in an atmosphere of pure hydrogen. The cupric solution (15 to 20 cubic centimètres) is placed in a small bottle with three necks, one for the admission the other for the exit of the gas, this latter being furnished with a small stirring apparatus, similar to that on the large bottle; on the middle neck of the bottle is fixed, by an india-rubber tube, the socket of the hyposulphite burette, lengthened and drawn out to a fine point, and charged beforehand. The reducing agent is allowed to run in as soon as the air is expelled, and this is continued till complete decolouration takes place; this point is somewhat difficult to ascertain.

Example.—By operating as above, we have found that—

(1) 4·6 cubic centimètres of hyposulphite were equivalent to 50 cubic centimètres of indigo.

(2) 15·1 cubic centimètres of hyposulphite were equivalent to 25 cubic centimètres of ammoniacal solution of copper; 10 cubic centimètres of decoloured cupric solution = 1 cubic centimètre of oxygen (at 0° C. (32° F.), and 760 millimètres of pressure). We easily find that 1 cubic centimètre of indigo, corresponds to 0·0152 cubic centimètres of oxygen.

On the other hand;

For 100 cubic centimètres of water to be tested, we have employed 5·7 cubic centimètres of some hyposulphite.

20 cubic centimètres of indigo require 2·9 cubic centimètres of this hyposulphite.

We make the proportion :—

$$2{\cdot}9 : 20 :: 5{\cdot}7 : x = \frac{20 \times 5{\cdot}7}{2{\cdot}9} = 39{\cdot}3 \text{ cubic centimètres.}$$

The oxygen of 100 cubic centimètres of water corresponds to 39·3 cubic centimètres of indigo. Nothing remains but to multiply 39·3 by 0·0152 = 0·59736, to find that one litre of water contained 5·97 cubic centimètres of dissolved oxygen, measured at 0° C. (32° F.) and at 760 millimètres pressure.

N.B.—It is important, when we make a series of analyses, never to use the last cubic centimètres of the burette of hyposulphite, which, having remained in contact with the air, have lost their reducing power. If we leave more than half an hour's interval between the trials, it would be better to renew entirely the contents of the burette.

A complete experiment, including the preparation of the reduced medium, does not require more than ten minutes, and each of the succeeding measurements is made in one or two minutes. Two trials of the same kind, repeated, never differ more than $\frac{1}{10}$ of a cubic centimètre.

This process of analysis allowing us to estimate the oxygen dissolved in 50 cubic centimètres of water (3·05 cub. in.) with an approximation of 0·005 cubic centimètre, and consequently of 0·1 cubic centimètre per litre, we have been able to utilize it for the purpose of studying the respiratory phenomena of yeast, and of measuring their intensity under different conditions of temperature. The rapidity of the estimates, which only require three or four minutes, gave us the opportunity of multiplying the experiments, and establishing the results which are given below by a series of analyses, the number of which was not obliged to be restricted. The experiments to which I shall here allude apply only to the case of fresh yeast in the form of paste, containing from 29 to 30 per cent. of solid matter, diffused in pure aerated water, without the addition of any nutritive element. We shall presently see what takes place when the yeast is placed under conditions more favourable to its developement.

The method consists in leaving a known weight of yeast for a known length of time in contact with a known weight of water, under the conditions of temperature which we may wish to study. The oxymetrical degrees of the water are measured at the commencement and the end of the experiment. Their difference gives the oxygen absorbed.

The yeast of beer only shows the phenomenon of absorption of oxygen, with production of carbon dioxide. All things else being equal, the respiratory intensity is the same in the dark, in diffused, and in full light; it is proportionate to the weight of yeast employed.

The original amount of dissolved oxygen does not sensibly influence the results, except when it is below 1 cubic centimètre per litre. We find, in this case, a feeble diminution in the rapidity of the absorption, which continues till complete deoxygenation of the water.

Below 10° C. (50° F.), the absorbing power of the yeast for oxygen is almost nil; it increases slowly to 18° C. (about 65° F.); from this point the increase is rapid till about 35° C. (95° F.), a temperature at which the respiratory intensity attains its maximum, which is sustained sensibly till 50° C. (122° F.). At 60° C. (140° F.), the absorbing power is annulled and destroyed.

A specimen of yeast, sensibly fresh, containing 26 per cent. of solid matter, absorbed, per gramme and per hour, at 9° C. (about 49° F.), 0·14 cubic centimètre; at 11° (52° F.), 0·42 cubic centimètre; at 22° C. (about 72° F.), 1·2 cubic centimètre; at 33° C. (about 93° F.), 2·1 cubic centimètre; at 40° C. (104° F.), 2·06 cubic centimètres; at 50° C. (122° F.), 2·4 cubic centimètres; at 60° C. (140° F.), none.

Another specimen of yeast, very fresh, and of very good appearance, containing 30 per cent. of solid matter, absorbed, per gramme and per hour, at 24° C. (about 76° F.), 2·2 cubic centimètres of oxygen; at 36° C. (about 97° F.) 10·7 cubic centimètres.

The increase of absorbing power between 24° C. and 36° C. was therefore more considerable than with the first yeast; this power is doubled in the one case, and quintupled in the other.

The values of the quantities of the oxygen absorbed, as well as the magnitude of the variations with relation to the temperature, are by no means absolute; they depend on a particular factor, inherent in the yeast, which we may call the factor of vitality; but whatever this factor may be, the direction of the variations is always the same, and susceptible of being represented by a curve starting from the line of abscissæ, or line of temperature, at about 9° or 10° C. (about 40° or 50° F.), rising slowly up to 18° C. (about 65° F.), thence rapidly reaching, at about 35° C. (95° F.), a maximum height, which it retains till nearly 60° C. (140° F.), when it returns suddenly towards the line of abscissæ.

Yeast can not only utilize and cause to disappear the oxygen physically dissolved in water, but also oxygen combined with hæmaglobin, which, as we know, can be eliminated by a diminution of pressure.

Thus, when we diffuse fresh yeast, whether washed or not, in arterial red blood, or in a solution of hæmaglobin saturated with oxygen, we see the tint change rapidly from red to dark blue or black. A simple agitation of the blood with air is sufficient to restore its ruddy colour; then the phenomena of deoxygenation recommence; the same experiment may thus be repeated a great number of times, especially with fresh and washed yeast.

Although, in this case, the yeast is in contact with a medium infinitely more rich in oxygen than is aërated water (containing from 200 to 230 cubic centimètres of oxygen to the litre, instead of from 6 to 7 cubic centimètres), the rapidity of the absorption is not increased, if the conditions of temperature are the same.

One gramme of yeast absorbs as much oxygen in an hour, at the same temperature, whether it be mixed with water containing 5 or 7 cubic centimètres of oxygen per litre, or in arterial blood containing 200 cubic centimètres of oxygen.

In the experiment with blood, we might fear a direct influence of the yeast, or of its soluble materials, on the colouring matter of blood; this influence is, in fact, produced, especially with solutions of hæmaglobin; it shows itself by the transformation of this primordial colouring matter into hæmatin; but it only makes its appearance after some hours.

The behaviour of the yeast with reference to blood may be explained in the following manner: The cells of *Saccharomyces* diffused in the liquid breathe at the expense of the oxygen physically dissolved in the plasm or serum in the midst of which swim the red globules of blood. In proportion as the plasmic liquid grows less rich in oxygen, a portion of this body, feebly combined with hæmaglobin, is separated, and enters into physical dissolution, by a dissociation comparable to that presented by potassium bicarbonate in a vacuum; the process continues till there is a complete disappearance of the oxygen dissolved in the serum, and of that which is fixed in the hæmaglobin. If this explanation is correct, the experiment ought to succeed, even when the blood is separated from the yeast diffused in water or serum by means of a membrane permeable to gas and to liquids, but capable of preventing all direct contact between the yeast-cells and the red globules. This is, in fact, what takes place. I have thus been able, by arranging a suitable apparatus, to imitate artificially that which takes place in the organs and tissues of animals, when the red and oxygenated arterial blood traverses the network of capillary vessels, and passes out into the veins under the form of black and partially deoxygenated blood.

For this purpose, it is only necessary to cause red blood to circulate slowly through a sufficiently long system of hollow tubes, the walls of which are formed of thin gold-beater's skin, which is immersed in a mixture of yeast diffused in fresh serum, without globules, kept at 35° C. (95° F.).

We see the red blood pass out black and venous at

the other extremity. A confirmatory experiment, made at the same time, with a system of tubes precisely similar, but immersed in serum without yeast, proves that yeast is indispensable for thus rapidly effecting the deoxidation of the blood. This experiment is the exact representation of what takes place in the animal organism, with the exception of the perfect method employed by nature to multiply contacts and surfaces.

In the latter case, the cellular and histological elements of the tissues play the part of the yeast; they absorb the oxygen dissolved in the plasmic liquids which bathe them, and constantly tend to bring down to zero their oxymetric condition. The oxgen, but feebly fixed in the hæmaglobin, re-establishes the equilibrium by a series of gaseous diffusions from the red globules to the plasm of the blood, and from the plasm of the blood to that of the organs. These continual diffusions are the inevitable consequence of the disturbance of equilibrium produced by the aëration of the organic cells, or of the cells of yeast in the experiment just described.

All these facts, then, prove distinctly that yeast breathes when placed in contact with dissolved oxygen The measure of the respiratory power, under the most favourable conditions, shows us this respiration to be as active, and even more so, than that of fishes. This cannot be considered as only a curious accessory fact, of which we must take but slight notice, in the study of the biological phenomena of yeast. For, *à priori*, it is improbable that a function so distinct, so sharply defined, is of no serious importance. On the other hand, if we consider what is passing in other living beings, from the highest to the lowest ranks in the scale of animal and vegetable life, we see respiration—that is to say, combustion—at the expense of oxygen, playing a preponderating part.

Without dwelling on the animal kingdom, we may remember that it has been long known that plants, in darkness, absorb oxygen, and disengage carbon dioxide; it was even suspected, with good reason, that this respiratory function, the reverse of that shown by parts exposed to the sun, was independent of the diurnal respiration, and that it belonged to another class of cells containing no chlorophyll. By operating on immersed aquatic plants, we can demonstrate this fact in the clearest and most elegant manner. Let the fresh stalks of Elodea be immersed in aërated water, of which the original oxymetrical degree is known. The flask, completely filled, is placed in the dark. At the end of two or three hours, we test the quantity of oxygen, and we find a diminution, which, as was the case with the yeast, at equal temperatures, is proportional to the quantity of the plant, and the duration of the experiment, and whose absolute amount varies with the temperature.

If, now, we warm for a moment the water and the plant to about 50° C. (122° F.), we shall destroy for ever the activity of these cells containing chlorophyll, that is to say, the powers which the plant possesses of decomposing carbon dioxide by its green parts; but without lessening its activity in respiration or combustion. We have seen, in fact, in yeast, that this function is only finally modified at about 60° C. (140° F.). The green parts of the plant are dead, but it is still capable of fulfilling certain biological functions. The flask of aërated water may be exposed to the sun's rays; when, far from observing an increase in the quantity of dissolved oxygen, the opposite is seen.

It is especially in the parts of plants which have to undergo a rapid evolution, and a marked cellular development, that the absorption of oxygen and internal combustion show themselves to be very active.

Every one knows that in the germination of seed or grain, that of barley in the manufacture of beer, for instance, the internal combustion develops a considerable quantity of heat.

The blossoming of flowers is also accompanied by a very marked oxidizing respiration.

Returning to yeast, M. Pasteur has shown (Bullet, Soc. Chimique, p. 80, 1861) that beer-yeast, sown in an albuminous liquid, such as the water of yeast, multiplies, even when there is no trace of sugar in the liquor, provided always that the oxygen of the air be present in large quantity; deprived of air, and under these conditions, the yeast throws out no buds. The same experiments may be repeated with an albuminous liquid mixed with a solution of unfermentable sugar such as crystallized sugar of milk; the results are of the same kind.

The yeast formed thus, in the absence of sugar, has not changed its nature, it causes sugar to ferment if it has been made to react on that substance without the access of air. We must, however, remark, that the development of yeast is very difficult when it has no fermentable matter to nourish it.

On the other hand, this same observer has remarked that, when in contact with air, and when it is extended over a large surface, alcoholic fermentation is more rapid than when deprived of oxygen, and that the budding is more active, since, notwithstanding the great rapidity of the fermentation, the relation between the newly-formed yeast and the decomposed sugar passes from $\frac{1}{80}$ to $\frac{1}{4}$ or $\frac{1}{10}$.

M. Mayer (Landw. Versuchs., vol. 16, p. 290) performed experiments, from which it would appear that oxygen has no influence, either on the rapidity of the fermentation or on the quantity of newly-formed yeast. However, his process of aëration of the liquids in fermentation, which consists in allowing calcined air to pass into the flask three times a day, appears to me to be insufficient, the slowness with which the water absorbs oxygen, and the rapidity with which yeast absorbs it, being well known; we cannot, then, draw from the experiments of this author the conclusions unfavourable to Pasteur's theory which he drew from them.

The results of all these facts are,—That yeast, like ordinary plants, buds and multiplies even in the absence of fermentable sugar, when it is furnished with free oxygen; that this multiplication, however, is favoured by the presence of sugar, which is a more appropriate element than non-fermentable hydrocarbon compounds; and also, that yeast is able to bud and multiply in the absence of free oxygen, but that in this case a fermentable substance is indispensable.

We are, therefore, compelled to arrive at the conclusion which M. Pasteur has drawn from all these facts; that saccharine matter can supply free oxygen in proportion to the yeast, and can excite it to bud. M. Pasteur has gone farther; he thinks that the fermenting character of a cell is due to the power which it possesses of breathing at the expense of sugar, without the contact of air, and that the decomposition into alcohol and carbon dioxide is the consequence of a disturbance of equilibrium, due to this partial abstraction of oxygen. We may also interpret the facts in the following manner:—

(1) The supply of oxygen, and the combustions to which it gives rise are necessary for the development and the reproduction of cell life. This fact is abundantly established for all the forms of life and organs of the vegetable kingdom.

(2) Yeast possesses the power of resolving the sugar which penetrates by endosmose into the interior of the cell into alcohol, carbon dioxide, glycerin, succinic acid, and oxygen. In fact, we have before seen (p. 23 of the French) that M. Monoyer had proposed a very simple equation to represent M. Pasteur's formula relative to the formation of succinic acid and glycerin.

In this equation which we give below,—

$$4\,(C_{12}\,H_{11}\,O_{11} + H\,O), \text{ or*}$$
$$4\,(C_{12}\,H_{12}\,O_{12}) + 6\,H\,O = C_8\,H_6\,O_8 + 6\,C_6\,H_8\,O_6 + 2\,C_2\,O_4 + O_2$$

we see an excess of oxygen make its appearance in the second member of the equation, and M. Monoyer adds, "we may suppose that this oxygen in excess serves for the respiration of the yeast. After this, the idea is not without foundation that fermentation is a primary phenomenon, due to a special action of the yeast and of other cells (Sechartier and Bellamy), and that,

* This is the old notation form of =" given by M. Schützenberger; the new notation form will be found on the page referred to.

as a consequence of this fermentation, there is some oxygen to spare, which may serve the purposes of respiration, and consequently may promote the budding of the yeast.

In this manner of explaining the facts, the yeast would not become a ferment because it breathes a part of the oxygen of the sugar ; but it may breathe a part of the oxygen of the sugar, and consequently reproduce, precisely because it sets free oxygen by decomposing the sugar.

Looking at the question in a more general point of view, we may also say that respiratory combustion is to the living organism a source of energy necessary for its development. In the decomposition of sugar there would be, according to the calculations of M. Berthelot, a disengagement of heat ; the quantity of heat set free in this phenomenon would be about $\frac{1}{15}$ of the heat disengaged by the complete combustion of the sugar decomposed. (Comp. Rend., vol. 59, p. 901.) In this estimate no account has been taken of the heat of solution of the sugar which disappears, nor of that of the solution of the alcohol formed, positive quantities which would tend to raise the heat of fermentation. It is not even necessary, therefore, to have recourse to the hypothesis of a combustion at the expense of the oxygen of the sugar, to explain how the phenomenon of fermentation can feed the combustion, and become a source of the energy indispensable to the development of the plant.

However this may be, there is an evident correlation, as M. Pasteur observes, beween fermentation and the development, nutrition, and budding of yeast. In fermentation without oxygen, the relation between the new yeast and the decomposed sugar will be necessarily greater than in fermentation with oxygen, since in the former case the budding takes place only under the influence of the oxygen furnished by the sugar, and in the second, of that furnished both by the medium and the sugar.

We ought to examine what takes place when yeast is left to itself, without nourishment and in a damp state, without the intervention either of saccharine matter or oxygen, before we proceed to the study of the chemical modifications produced in yeast, while it is placed under conditions favourable to its nutrition and development. These conditions, as we have seen, return always to those of alcoholic fermentation, sugar being one of the indispensable sources of nourishment of the *Saccharomyces*, and this cannot be in contact with it without fermenting, provided that the other conditions of nutrition are fulfilled. Does it preserve, in this case, its original integrity, without any change in the chemical composition of its immediate principles? In other terms, does its vitality remain in a latent state, to manifest itself afresh, as soon as sugar or oxygen is supplied? *A priori*, this would appear improbable, if we regard what takes place in the tissues of plants. In fact, experiment has proved that, under these conditions, the yeast undergoes important modifications, in respect of the composition of its organic principles.

This question has been studied by M. Pasteur first, and then successively by M. Béchamp and by the author of this book.

M. Pasteur having set 5 grammes of sugar to ferment with 10 grammes of yeast, in the form of paste, (2·155 grammes of dry matter), a much larger quantity than is necessary for the complete decomposition of the sugar, was surprised to see that this fermentation did not entirely end, that it had a tendency to continue with a weak disengagement of gas, and when Fehling's liquor no longer revealed in it the slightest trace of sugar. Having arranged in receivers reversed over mercury the following fermentations:—

1st.	Sugar-candy	1·313
	Yeast from wine (deposit in empty casks)	6·950
	Pure water	9·336
2nd.	Sugar-candy	1·4425
	Yeast from beer (2·15 grammes when dry)	10·0
	Pure water	9·210

He obtained, in two days, while the gaseous disengagement was still sensible, from—

No. 1. 360 cubic centimètres (21·96 cub. in.) at zero (32° F.), and at 760 millimètres' pressure;

No. 2. 387·5 cubic centimètres (23·6 cub. in.) at the same temperature and pressure.

Of carbon dioxide, entirely absorbable by potass. The theoretical quantities, even when a deduction is made of the succinic acid and glycerin—that is to say, those which correspond to Gay-Lussac's former equation—those which give the highest number for carbon dioxide would be—

No. 1.	341·8.
No. 2.	375·5.

By increasing still more the amount of yeast, we arrive at more conclusive results.

Thus, ·424 gramme of pure sugar-candy, made to ferment with a weight of damp yeast corresponding with 10 grammes of dry matter, furnished, at the end of two days, 300 cubic centimètres of carbon dioxide; the sugar alone could only furnish 110 cubic centimètres.

The liquid, carefully distilled, gave a little more than 0·6 gramme of absolute alcohol. The weight of alcohol obtained was greater than the whole weight of the sugar employed, and in proportion to the volume of carbon dioxide formed.

This experiment proves that when yeast is mixed with a comparatively very small weight of sugar, after the latter has been decomposed, the activity of the yeast continues, reacting on its own tissues and its hydrocarbons with extraordinary energy and rapidity, proceeding more and more slowly as the process goes on.

If we put an end to the fermentation at the moment when a volume of carbon dioxide is formed, equal or very little greater than that which corresponds with the weight of the sugar employed, *we find no more sugar in the liquor.* This observation is very important, because it tends to prove that the action which the yeast exerts on its own elements does not commence till it is deprived of sugar.

It is not necessary to fulfil the conditions of the preceding experiments (fermentation with excess of yeast), in order to observe fermentation at the expense of the elements of the yeast itself; it is sufficient, as Pasteur has already shown, to mix fresh beer-yeast with water at 25° C. (77° F.); we soon see numerous bubbles of pure carbon dioxide gas rise, and it is easy, by distillation, to ascertain the production of alcohol. Hydrogen gas and signs of putrefaction do not appear till long afterwards, when the microscope reveals the presence of lactic ferment and infusoria.

M. Béchamp, who took up this question after M. Pasteur, took care to avoid completely the formation of infusoria, by employing creosote water; he also ascertained that alcohol and carbon dioxide were produced as a consequence of the vital activity of the yeast when without nourishment (deprived of oxygen and sugar). But this is not all, for this curious phenomenon is allied to other chemical reactions not less interesting.

It is well known that, by preserving yeast, in the form of damp paste, in a hot place (25° to 30° C., 77° to 86° F.), it undergoes considerable softening, and entirely changes its appearance.

This modification is not due to incipient putrefaction. There is no foundation for this conclusion; nothing is formed as a volatile product, except carbon dioxide and alcohol. The microscope reveals no organism except the *Saccharomyces*, especially if, as M. Béchamp suggests, we make use of creosote water. If, now, we treat this softened yeast with luke-warm water, we can extract from it a much greater quantity of soluble and diffusible principles than by employing fresh yeast.

Thus, 100 grammes of fresh yeast, leaving after dessication at 100° C. (212° F.), a fixed residue of 30 grammes, give after washing, before the softening takes place, only 23 grammes of dry residue. The loss by washing is 8 per cent.

The same yeast, softened spontaneously for two days, and then washed, leaves a residue weighing, when dry, 14 grammes; the loss by washing is thus 16 grammes.

Yeast has therefore, by softening, transformed into soluble principles 8 grammes of previously insoluble principles, per cent. of damp yeast, or 26·66 grammes per cent. of dry yeast. During this internal reaction of yeast, when kept damp, and without nourishment, M. Béchamp observed the production of pure nitrogen.

The water by which it has been washed contains acetic acid, a perceptible portion of soluble alterative ferment (zymase, of which we shall presently speak; *see* the chapter on soluble ferments), an albuminoid principle, soluble, coagulable by heat, and nearly allied to albumin, from which it differs by rotating power; a gummy matter, which nitric acid transforms into mucic acid, very like arabin, from which, however, it differs by rotating power; we find, also, tyrosine and leucine, and an uncrystallizable sirupy matter, alkaline and alkaline-earthy phosphates in observable proportions.

Resuming the question, the author confirmed the results obtained by M. Béchamp, and added to them new facts. The extract of softened yeast contains, besides the principles mentioned above, nitrogenous compounds of the sarcine group, which had not hitherto been noticed in the vegetable economy.

The following is the analytical process which I have followed. The yeast, digested without the addition of any nutritive principles, is boiled with a considerable quantity of water to coagulate the albumin; it is then filtered. The liquid, which has a slightly acid reaction, is concentrated in a water bath to a sirupy consistence;

it assumes, as it grows cold, a gelatinous crystallized form composed of small crystals forming a kind of paste in a brownish syrup. The whole, placed in a flask, is boiled for some time with a great excess of strong alcohol (92 per cent.); a deep coloured pitchy mass is separated, which sticks to the sides of the flask. The alcoholic solution, when properly concentrated, furnishes, as it grows cool, an abundant crystalline deposit, which, after filtration, washing with cold alcohol, and pressure, is almost white. This crystallized mass, formed of very thin flakes or hyaline globules, as well as that which is obtained by concentrating the alcoholic mother-liquor, is almost exclusively formed of sulphuretted *pseudo-leucine*, with a little tyrosine. The mother-liquor separated at the second crystallization is distilled in a water-bath to drive off the alcohol. The remainder is diluted with water, and solution of barytes is added to precipitate the phosphates. The excess of barytes is removed from the filtered liquid by a current of carbon dioxide. The filtered liquid is then boiled with an excess of cupric acetate. A brownish flaky precipitate is formed, which contains carnine, sarcine, xanthine, and guanine, combined with copper oxide. The filtered liquor above this precipitate, which is comparatively not very abundant, yields, when the copper has been removed by sulphuretted hydrogen, and it has been concentrated, a crystalline mass, from which cold alcohol extracts a sirupy nitrogenous substance, of a sweet taste, leaving a crystallized mixture of leucine and butalinine.

The cupric precipitate, produced by boiling in cupric acetate, is thoroughly washed with hot water, then treated, by the warm process, with dilute hydrochloric acid. Nearly all of it is dissolved, with the exception of black flakes of copper sulphuret. The solution, filtered while warm, deposits, when cold, a large part of the copper combination, which had been dissolved.

This deposit, washed and decomposed by sulphuretted hydrogen, furnishes carnine, which is purified by being crystallized in water, being deprived of colour, if necessary, by a little animal charcoal. The hydrochloric mother-liquor which deposited the copper combination of carnine, having been deprived of its copper by sulphuretted hydrogen, and then concentrated, gives first crystals of xanthine hydrochlorate ; then, by concentrating the decanted liquor in the cold, crystals of guanine hydrochlorate, from which the guanine is extracted by precipitating it with an excess of ammonia, which dissolves the xanthine, and leaves the guanine.

The sarcine is obtained by precipitating by ammonia and silver nitrate the nitric solution of the first copper precipitate ; by washing with water of ammonia the dirty-white gelatinous precipitate which is formed, and then crystallizing it in boiling nitric acid at 12° Baumé ; we thus obtain immediately the nitro-argentine combination of sarcine. This is decomposed by sulphuretted hydrogen ; the liquid, when filtered, and concentrated, has ammonia added to it, which, by concentration, leaves the sarcine in fineneedle-like crystals.

The pitchy precipitate formed at first, by the addition of alcohol to the concentrated extract of digested yeast, is in great part formed of earthy phosphates, tyrosine, and gummy matter.

The leucine extracted from the yeast after its digestion shows a peculiarity noticed by Hesse. It contains a quantity of sulphur, which varies within certain limits, and may attain to 4 per cent. My analyses, made from very pure products, crystallized several times in alcohol, and having the appearance of beautiful white nacreous flakes, gave from 2·93 to 2·14 per cent. of sulphur. This cannot be eliminated by long boiling at 100° C. (212° F.) with a mixture of potass and lead hydrate. The hydrogen of these leucines have always been found to be somewhat deficient (9·34 or 9·6 per cent. instead of 9·9).

It appears that the sulphur forms an integral part of the molecule, and is not found in the state of a sulphuretted body mixed with leucine. At any rate, repeated and very careful purifications do not succeed in splitting up this mixture.

Let us now see what may be the signification of these results.

All the nitrogenous compounds noticed as being the products of spontaneous changes in the yeast have been obtained directly by the splitting up of albumin and albuminoid substances. (*See* the chapter on these bodies.) Their origin is, therefore, not doubtful; they are formed by the decomposition of certain insoluble proteids in the yeast, and by a chemical process similar to that which takes place in the animal tissues; for it is impossible to mistake the great analogy of composition which exists between the extract of digested yeast and extracts obtained from animal tissues. This chemical phenomenon is also comparable to those which are observed by causing dilute warm sulphuric acid, or the alkalis, such as potash and barytes, to react, under certain conditions, on proteids; we thus obtain, in fact, similar products.

Albumin, obtained in such large proportions in the extract of digested yeast, must be one of the results of the splitting up; it is probable that yeast has no action on the one substance among proteids which resists so strongly chemical agents.

This opinion is corroborated by the interesting observation of M. Gautier, who proved that fibrin splits up, under the influence of salt water, into albumin and another albuminoid principle. Zymase, or alterative ferment, which exists in large proportions in the extract of softened yeast, also represents one of the substances derived from the decomposition of insoluble proteids.

I have before said that we succeed, by direct analysis, in separating an uncrystallizable nitrogenous principle of a sweet taste. This substance resembles in its properties the hemiproteidin on hemialbumin formed by the action of boiling dilute sulphuric acid on albumin.

As to the tyrosine, leucine, butalanine, the sarcinic bases, is evident that they are the directly derived products of albuminoid substances.

With respect to the proceeds of the sugar which furnish the alcohol and carbon dioxide of the spontaneous fermentation of yeast, and of the gummy matter, their source has not been ascertained with as great certainty.

It is generally admitted, in accordance with the views of Payen and Schlossberger, that washed yeast is composed of cellulose and insoluble proteids.

The question whether the sugar and gum which show themselves in notable proportions in the extract of yeast, when digested and washed, ought to be considered as the products of the physiological decomposition of the albuminoid substances, or the results of a transformation of cellulose, can be decided only by a series of quantitative experiments. The following results give only an approximate solution, but they prove, at least, that the greater part, if not the whole, of the solid principles contained in the extract of yeast when washed in cold water and digested, must be derived from proteids.

100 grammes of fresh yeast contain 30 grammes of solid matter. This includes 9·28 per cent. of nitrogen; whence it results, that 100 grammes of fresh yeast contain 2·78 grammes of nitrogen.

100 grammes of fresh yeast, washed with boiling water until it runs off clear, leave 20 grammes of solid matter. This contains 10·17 per cent. of nitrogen. The result is that 100 grammes of yeast, washed with boiling water, contain 2·03 grammes of nitrogen.

100 grammes of yeast, digested for fifteen hours at 35° C. (95° F.), and then washed with boiling water, contain 12·5 grammes of solid matter, yielding 7·55 per cent of nitrogen.

100 grammes of yeast, digested and then washed with boiling water, contain, then, 70·94 grammes of nitrogen. The loss of nitrogen arising from the digestion and washing is 1·82 – 0·75 = 1·07.

The loss of solid products due to digestion and subsequent washing amounts to 17·5 – 10 = 7·5.

But proteids contain, on an average, 15·5 per cent. of nitrogen; 1·07 gramme of nitrogen correspond to 6·9 grammes of proteinic matter. Consequently, out of 7·5 grammes of the principles which have become soluble by the process of digestion, 6·9 grammes or $\frac{14}{15}$ are derived from albuminoid substances.

On the other hand, a direct quantitative estimate of

the nitrogen in the dry extracts gave 12·5 per cent. of nitrogen. It is, then, evident that the yeast must have given up part of its non-nitrogenous elements; this last calculation would lead us to the proportion of $\frac{4}{5}$ proteids, and $\frac{1}{5}$ hydrocarbon matter, if we take no account of the elements of water which are necessarily united with the albuminoid matter at the time of its splitting up and transformation into bodies such as leucine. By making an approximate estimate of this water of hydratation, we should find $\frac{10}{11}$ of albuminoid and $\frac{1}{11}$ of hydrocarbon matter. It is not very probable that the gum has an albuminoid origin; it could at most be derived only from the decomposition of a substance analogous to tunicin or chitin.

The principles yielded to water by fresh yeast in a much smaller proportion are of the same nature as those which we have found before; it cannot, indeed, be otherwise. The conditions under which M. Béchamp and I carried on our researches necessarily exaggerated the effects of a continuous cause. As soon as the yeast finds no more sugar, it reacts on its own elements, and it is difficult to find yeast which has not been placed in this situation for a longer or shorter time. It is probable that all kinds of yeast give reactions of this nature in the fermenting vats, when the sugar begins to fail. M. Béchamp says that he has found tyrosine and leucine in the aqueous extract of all fermentations which he has examined *ad hoc*. Before, therefore, we draw too positive conclusions, we must know whether the fermentations have been studied immediately after the total decomposition of the sugar.

M. Pasteur has shown us that spontaneous fermentation does not appear till this moment; it would be interesting to ascertain whether the formation of nitrogenous excremental substances, such as leucine or tyrosine, takes place during fermentation or no; we should thus see whether this phenomenon is allied to spontaneous fermentation, or is independent of it.

M. Béchamp, considering the spontaneous fermentation of yeast and its concomitant phenomena (the production of acetic acid, tyrosine, &c.) has given a physiological theory of fermentation which may be shortly stated thus:—

Yeast, like every living organism, shows phenomena of two kinds; those of nutrition and assimilation, which are subordinate to the presence of its nutritious principles (sugar, nitrogenous compounds, mineral salts). These various principles, penetrating by endosmose into the cell, undergo there suitable transformations, and are converted into tissues of recent formation in the new cells which are formed by budding. Together with these phenomena of nutrition, and side by side with them, other inverse reactions, those of disassimilation, take place, by which the tissues are changed into excrementitious products, unsuited to the life of the cell, and these are eliminated. The production of carbon dioxide and of alcohol are the consequences of this process, and belong to disassimilating reactions. In this theory, M. Béchamp develops the ideas of M. Dumas as to the part played by yeast and other ferments. It is certain that the production of carbon dioxide and alcohol without sugar, at the expense of the constituent elements of the yeast itself, gives some support to the opinion of Béchamp, whether or no we admit that this production is the result of the formation of buds, in which the new cells are nourished at the expense of the old ones.

On the whole, this theory throws but little light on the question relating to the very essence of the phenomenon. Whether the sugar is decomposed with or without previous assimilation is of little importance; we are no nearer knowing why it is decomposed, and scarcely any one now doubts that the decomposition of sugar is a biological phenomenon. We must now examine some special points relating to alcoholic fermentation.

It has been long thought that alcoholic fermentation could take place under two distinct circumstances, according as yeast is added to a solution of pure sugar in water, or to sugared water containing the nitrogenous and saline principles necessary for its nourishment. In the first case, it was thought that the ferment acts without reproducing itself; while in the second case, which is that occurring in the brewing of beer, it acts and reproduces itself. More than this, Thénard had observed that, in the fermentation of 20 parts of yeast and 100 parts of sugar, there only remain, after all the disengagement of carbon dioxide has ceased, 13·7 parts of an insoluble residue, which may be reduced even to 10 by fresh contact with sugar. Thus yeast, while exciting the fermentation of pure sugar, partially destroys itself.

M. Pasteur, on the contrary, admitting it to be proved by his experiments that the budding and multiplication of yeast are phenomena which, in a constant manner, accompany all alcoholic fermentation; explains the formation of new globules in a solution of pure sugar in water by nutrition at the expense of the nitrogenous soluble substances of the original yeast. In this case, whatever soluble nitrogenous aliment there is in the ferment employed becomes fixed in an insoluble state in the newly-formed globules. It is certain that yeast formed in a suitable medium, such as the wort of beer, is gorged with soluble nitrogenous principles, which it can yield to water, and which are eminently suited to the nutrition of new yeast.

In order to remove all doubt as to the reality of this explanation, it was necessary for Pasteur thoroughly to explain away those of Thénard's results, quoted above, which seem to contradict his opinion.

The following table sums up his observations on the fermentation of pure sugar with yeast :—

	Weight of pure sugar-candy.	Weight of yeast, washed while fresh, in the state of a soft paste.	Weight of the yeast when dried at 100° C. (212° F.).	Weight of yeast deposited after fermentation, when dried at 100° C. (212° F.).	Weight of extract, the soluble part of the yeast remaining in the fermented liquid, and insoluble in the mixture of alcohol and ether.	Sum of the weights of the yeast deposited after fermentation and of the soluble extractive part remaining in the fermented liquor.	Excess of this sum over the weight of yeast exposed to fermentation.
	gr.	gr.	gr.	gr.	gr.	gr.	gr.
A	100	20·000	4·626	3·230	2·320	5·550	0·934
B	50	10·000	2·213	2·001	0·819	2·820	0·407
C	100	16·000	4·604	4·385	not determined.	—	—
D	100	10·000	2·313	2·486	1·080	3·566	1·253
E	100	13·700	2·626	2·965	0·964	3·929	1·303
F	100	6·254	1·198	1·700	0·631	1·331	1·133
G	16	3·159	0·699	0·712	not determined.	—	—
H	4	1·474	0·326	0·335	—	—	—
I	20	1·878	0·476	0·590	0·133	0·723	0·247

We see that in the experiments A B C, in which the weights of yeast in the form of paste is above 15 or 20 per cent. of the weight of sugar, after fermentation less yeast is collected than is originally put in; these are, therefore, precisely the conditions of Thénard's experiments, who employed and recommended 20 parts of yeast in form of paste to 100 of sugar.

If, on the contrary, the weight of yeast in the form of paste is equal to or less than 10 per cent. of the sugar, more yeast is collected than had been first employed (exper. D E F G H I).

And, in every case, by adding the weight of extractive matter to that of the yeast remaining after fermentation (having deducted the glycerin and succinic acid which are in solution in the fermented liquor, and come from the yeast), we find that the sum always exceeds very sensibly the whole weight of the original yeast.

The increase amounts to about 1·2 or 1·5 per cent. of the weight of the sugar decomposed.

The result of this appears to be, that in Thénard's experiments, the disappearance of yeast was only apparent. Less was collected because much was used, and because the soluble part which passes into the liquid is greater than the weight of the newly-formed globules.

M. Duclaux (Thèses de la Faculté des Sciences de Paris, 1865) arrived at results of the same kind.

In 100 grammes of fermented sugar he found—

Dry yeast employed.	Dry yeast obtained.
17·32	14·02
8·66	8·51
4·33	4·78
2·16	2·58

However, in other trials made by the same author, the weight of yeast found exceeded that of the yeast employed, even in the case in which the original proportion exceeded 16 per cent. of sugar. This difference is attributed by the author to the use at first of a kind of yeast poorer in extractive principles, and yielding fewer solubler substances to the water.

The results obtained by Pasteur and Duclaux, in fermentation conducted on a small scale, have been confirmed by other experimentalists, such as those of Mayer, Fitz.

From the whole results published, it seems that we may conclude that, within certain limits, and with some variations, the weight of the newly-formed yeast is in proportion to the weight of the sugar decomposed, and equal to from 1·5 to 3 per cent. of the fermented sugar.

We ought, however, to add, that the information afforded by commercial experience, by the manufacturers of compressed yeast, do not agree with these conclusions.

According to Märcher and Schulze, we are able to obtain 28·6 parts of dry yeast for 100 of alcohol, or, which comes to the same thing, 14·66 parts of dried yeast for 100 parts of decomposed sugar.

In the fermentation of the must of grapes, the formation of yeast seems much greater in proportion to the fermented sugar than M. Pasteur's ratio would make it.

M. Pasteur, who carries on with so much success the manufacture of pure yeast on a large scale, by the fermentation of the wort of beer in a close vessel, obtains 1·5 kilogrammes (3·3 lbs. avoir.) of expressed yeast, which contains 30 per cent. of dry matter, say 450 grammes

('99 lbs. avoir.) of dried yeast per hectolitre (22 gallons) of beer; allowing that this beer contains 8 per cent. of alcohol, fermentation would have acted on 15'5 kilogrammes of sugar, there would then be 2'9 grammes of new yeast for 100 of decomposed sugar; these numbers do not differ so much as the first, from the results obtained on a small scale; it is true, however, that in this calculation, we have taken for the proportion of alcohol a very high weight which ought, perhaps, to be reduced.

The apparent contradiction between the results of the laboratory experiments and those of commercial operations, may be explained by admitting, that the development and multiplication of the cells of yeast are not so much dependent on the quantity of sugar decomposed, as the laboratory experiments seem to indicate, that they may vary within very wide limits, according to the composition of the medium in which the operation takes place.

This explanation seems to be corroborated by the following experiments of Mayer (Landwi, Versuchsz, vol. 16, p. 304).

The author fermented two portions of the boiled must of grapes of the capacity of 190 cubic centimètres (11'59 cub. in). To one, he added imperceptible traces of wine-lees (*Saccharomyces ellips.*), and sowed in the other a little beer-yeast. When the sugar had disappeared, which required several weeks, he found—

	ALCOHOL FORMED.	CORRESPONDING SUGAR DECOMPOSED.	YEAST FORMED.
	gr.	gr.	gr.
1. Saccharomyces ellip. .	13'11	25'6	2'38
2. „ cerevisiæ	15'2	29'7	2'04

Therefore for 100 parts of sugar decomposed there were formed—

No. 1. 9·2 of yeast.
No. 2. 6·8 „

numbers much higher than the results obtained in fermentations under other conditions of the medium.

On the other hand, M. Pasteur found that in fermentations, in the presence of nitrogenous and mineral nutritive matter, nearly 1 per cent. of the weight of sugar was formed in yeast and soluble products; a little less, therefore, than when we operate with yeast completely formed, and pure sweetened water.

Thus, to 9·899 grammes of pure sugar-candy, dissolved in a sufficient quantity of water, were added 20 cubic centimètres of boiled and filtered water with which fresh yeast had been washed (containing 0·334 grammes of albuminoid and mineral matter), it was then filled up to 100 cubic centimètres and a trace of fresh yeast added. The sugar had disappeared at the end of eleven days; 0·152 grrammes of dry yeast were collected. The nitrogenous remainder left after the evaporation of the liquid, when the glycerin and succinic acid had been removed, weighed 0·260 grammes.

Thus, 0·334 grammes of nutritive matter were employed, and 0·152 grammes of yeast + 0·260 grammes of soluble nitromineral matter was found; in all, 0·412 grammes. The difference is 0·78 grammes.

When pure sugar ferments with a limited quantity of yeast, this becomes exhausted, and at last becomes unfit to effect fresh decompositions of sugar, in the absence of soluble nutritive materials. This ought, in fact, to be the case. The amount of the nutritive materials which are soluble, or become so during the fermentation, and which are contained in the limited yeast, being gradually utilized for the formation of fresh globules, there ought to come a time when the yeast will be destroyed, for want of nutriment. This phenomenon of the nutrition of the yeast at its own expense, does not exclude the presence of nitrogenous matter in the no longer fermenting saccharine liquid, for it may happen that a part of the nitrogenous principles eliminated by the yeast are unsuited to its nourishment, and only represent excremental products. Leucine and tyrosine seem to belong to this order of products; in fact, some direct experiments of Mayer have proved that they are unsuited to development.

In a fermentation, in which 1·198 grammes of washed yeast (the weight of dried matter, containing 9·77 grammes per cent. of nitrogen) had been employed to ferment 100 grammes of sugar, 1·745 grammes of yeast were collected after fermentation; this contained 5·5 per cent. of nitrogen; the extractive residuum of the fermented liquid, when deduction had been made of the succinic acid and glycerin, weighed 0·6 grammes and contained 3·8 per cent. of nitrogen. Under these conditions, the yeast was exhausted—that is to say, the soluble nitrogenous principles contained in the liquid had become unfit for the development of new cells, and represented excremental products. We know that among the products left undetermined by M. Pasteur, we must place leucine and tyrosine.

This experiment shows us the cause of the diminution of nitrogen in the yeast after fermentation. On the one hand, the total weight of yeast has increased by the addition of principles (cellulose) furnished by the sugar, and not nitrogenous; on the other hand, the original yeast has yielded to the liquid soluble nitrogenous products which have become unfit for its nutrition. Thus —the apparent disappearance or diminution of the nitrogen of the yeast during fermentation with pure sugar is explained in a very natural manner, a phenomenon which had so puzzled Thénard, and which Döbereiner had wrongly thought might be attributed to the formation of ammoniacal salts. We have already seen by Pasteur's very careful experiments, that the ammonia of the liquid has a tendency rather to disappear than to increase during the fermentation.

According to Dubrunfaut (Comp. Rend., July, 1871), the wort of beer, which, under ordinary circumstances, reproduces sevenfold the weight of the yeast employed, is so rich in matter reproductive of ferment, that it is far from being exhausted by this process. An addition of sugar excites an increase proportional to the sugar supplied. If it has not been employed in excess, the normal condition of the ferment with respect to nitrogen has not changed; if the contrary has been the case, the amount of nitrogen lies between that of fertile and of barren yeast. All kinds of yeast, the results of fermentation other than that of beer made from malt, are in this condition; they give an amount of nitrogen varying between 0·10 and ·05.

Beer-yeast, notwithstanding the loss of weight sustained in washing, preserves its proportion of nitrogen, while its richness in salts decreases from ·10 to ·02. However often it may be washed, we can never obtain water quite free from albuminoid and saline matter; which proves that the yeast continues to live, even in pure water, and to exercise there its vital functions on its own substance, as animals do which are condemned to starvation.

The ash derived from water in which yeast has been washed, is always alkaline; that of the washed yeast is acid; which induces us to believe that there is an ammonio-magnesian phosphate present in the yeast, which, in consequence of its insolubility, remains in the washed ferment. M. Dubrunfaut has, like M. Pasteur, ascertained the constant disappearance of a certain portion of the ammonia of the medium, as a concomitant of the reproduction of ferment; but at the same time, the ash becomes more acid. The addition of ammoniacal salts to fermentation going on under unfavourable conditions, has the effect of facilitating the fermentation, and preserving the proportion of nitrogen, 0·1 of the yeast.*

M. Dubrunfaut performed a series of experiments to ascertain the part played by different mineral salts in the fermentation of sugar, and in the development of yeast. He made use for this purpose of solutions of pure sugar at 10 per cent., and added to them various mineral salts and yeast in the form of paste, all taken at a weight 0·05 of the weight of the sugar. The ferment employed, therefore, only represented in dry matter 0·01 of the weight of sugar. He tried, 1. potassium nitrate; 2. ammonium sulphate; 3. potassium sulphate; 4. calcium phosphate; 5. magnesium sulphate; 6. calcium

* These last experiments would have been more properly placed in the chapter concerning the influence of salts and nitrogenous matter, on the development of yeast. We only give them here as a supplement, which the reader will be kind enough to read for himself in its proper place.

sulphate ; 7. sodium sulphate ; 8. a wort without mineral salts, to serve for comparison :—

	FRACTIONS OF SUGAR DECOMPOSED IN THE SAME TIME.
Sodium sulphate	0·52.
Calcium "	0·62.
Magnesium "	0·73.
Calcium phosphate	0·80.
Potassium sulphate	0·88.
Ammonium "	0·94.
Potassium nitrate	1·10

for complete and perfect transformation without production of acid.

All these salts gave higher results than those of the wort used for comparison, which only transformed 0·50 of the sugar into alcohol.

The nitric acid, in this experiment, entirely disappeared. It appears, then, from Dubrunfaut's observations, that yeast forms no exception with respect to the facility with which nitric acid is assimilated.

CHAPTER VI.

ACTION OF VARIOUS CHEMICAL AND PHYSICAL AGENTS ON ALCOHOLIC FERMENTATION.

IT has long been known that certain chemical compounds, especially those which coagulate albuminous substances and disorganize the tissues, or which, by their presence in sufficient quantities, are incompatible with life, are opposed to fermentation; such are the acids and alkalies in suitable proportions, silver nitrate, chlorine, iodine, the soluble iron, copper and lead, salts, tannin, phenol, creosote, chloroform, essence of mustard, alcohol when its strength is above 20 per cent., hydrocyanic and oxalic acids, even in very small quantities.

An excess of neutral alkaline salts or sugar acts in the same manner, by diminishing in the interior of the cell the minimum quantity of water which is necessary for the manifestation of its vital activity.

The red mercury oxide, calomel, manganese peroxide, the alkaline sulphites and sulphates, the essences of turpentine and of lemon, &c., also interfere with, and destroy, alcoholic fermentation.

Phosphoric and arsenious acids are, on the contrary, inactive. We owe to M. Dumas (Ann. Chim. Phys.,

5th series, vol. 3, 1874, p. 81) a very elaborate work on the question which we are now considering; we will give here the principal results. This illustrious chemist first proved, by many experiments, that the fermentation of sugar under the influence of yeast can be studied like any regular phenomenon, which, when subject to determinate disturbing forces, would be able to show their influence with precision.

Thus, at 24° C. (about 76° F.) 20 grammes of yeast decompose 1 gramme of sugar (glucose) in twenty-three or twenty-four minutes, (the mean of several trials agreeing with each other).

In another series of experiments it was found that 100 grammes of yeast decompose, at the same temperature, 1 gramme of glucose in twenty-four minutes. Thus, yeast acts on glucose with the same rapidity, in the proportion of 20 grammes of yeast to one of glucose, as of 100 to 1.

Experiments made on the same yeast, with cane-sugar and glucose separately, showed that the destruction of 1 gramme of cane-sugar by 40 grammes of yeast had a maximum duration of thirty-four minutes, while one gramme of glucose required but sixteen or seventeen minutes.

If we mix some beer-yeast with water, and add to similar portions of the liquid, containing, for example, 150 cubic centimètres of water and 10 grammes of yeast, quantities of sugar represented by 0·5 grammes, 1 gramme, 2 grammes, 4 grammes, we find that the time necessary for the destruction of the sugar is exactly proportional to its quantity; or, in other words, that under identical circumstances the duration of the fermentation is proportional to the quantity of sugar, it being understood that the yeast is in excess.

By taking, then, for the length of the co-ordinates measured along the axis of the abscissæ the quantities of sugar, and for the co-ordinates measured along axis of the ordinates, the number of minutes necessary for the disappearance of the sugar, we obtain a straight line.

M. Dumas estimates approximately, according to his results, that to decompose 1 gramme (15·4 grains) of sugar in one hour, it requires 400 milliards (400,000,000,000) cells, supposing them all to be active.

The fact discovered by Dumas, that beyond a certain limit an excess of yeast does not promote the fermentation of a given quantity of sugar, is extremely important.

In fact, if the decomposition of sugar is a biological phenomenon which takes place in the interior of each cell into which the sugar penetrates; if it ought to be attributed to the nutrition of the globules of yeast, it seems that the rapidity of the decomposition of the same quantity of sugar ought to increase indefinitely with the number of cells which are in action, in the same manner as in a meal, the quantity of food consumed increases with the number of guests.

Temperature.—We have to determine three data relating to alcoholic ferment, with respect to heat:—

1. What are the two limits, maximum and minimum, compatible with the life of the ferment? 2. At what temperature is alcoholic fermentation, that is to say, the activity of the ferment, the greatest? The lower limit

compatible with the life of the *Saccharomyces* is not yet determined with precision.

It is only towards 8° or 10° C. (about 46° to 50° F.) that the cell shows its chemical energy by sensible reactions (fermentation, absorption of oxygen, &c.), but it may be cooled with impunity towards zero (32° F.), and even below, if we are careful to subject it, after congelation, only to very slow progressive variations, to re-melt the frozen water in which it is contained. The higher limit depends on whether we operate on yeast previously dried or still damp.

Thus, yeast dried carefully may be heated to 100° (212° F.) without losing its vitality.

Damp yeast would lose its activity (Mayer) towards 53° C. (about 128° F.), or at least at a temperature less than 60° C. (140° F.).

As to the most favourable conditions of temperature for good fermentation, they appear to range between 25° and 30° C. (77° and 86° F.), varying within narrow limits, according to the nature of the medium and that of the ferment; even near zero (32° F.) the production of alcohol by yeast is not absolutely destroyed.

Electricity.—The sparks of a Holtz machine, or those of an induction coil passing through water containing yeast, modify neither its power of changing cane-sugar into glucose, nor its activity as an alcoholic ferment.

Gas-light.—The process of fermentation is more slow in the dark. M. Dumas placed beer-yeast of the consistence of thick broth in flasks respectively filled with oxygen, hydrogen, nitrogen, carbon oxide, nitrogen protoxide, proto-carburetted hydrogen. At the end of three days, this yeast, placed in contact with sugar, acted in the same way as yeast kept in the open air, and the microscope showed no change.

Alcoholic fermentation is more slow in a vacuum.

Sulphur.—The presence of flour of sulphur, even in equal proportions to that of the dried yeast, did not sensibly interfere with alcoholic fermentation, as had been predicted; only the carbon dioxide which was disengaged contained 1 or 2 per cent. of sulphuretted hydrogen.

Action of Acids. — Yeast is always acid. If we neutralize the free acid which is found in it by lime-water, we find that in about five minutes the acidity reappears. We are therefore obliged, in order to maintain a continuous neutrality, to add for each gramme of yeast under examination, a quantity of lime equivalent to ·003 grammes of normal sulphuric acid. M. Dumas wished to ascertain whether the acidity of yeast can be increased, without its power being impaired, and whether the specific nature of the acids exercises any influence on the result.

Experimenting on sulphuric, sulphurous, nitric, phosphoric, arsenous, boracic, acetic, oxalic, and tartaric acids, he added to a mixture of sugar, water, and yeast (5 grammes yeast, 10 grammes sugar, 25 grammes water) quantities of each of these acids respectively, first equivalent, then tenfold, and a hundredfold of the acid power of the yeast.

The addition of one of these acids, even in feeble proportions, neither hastened the commencement nor the end of the fermentation. It often arrested the decomposition of the sugar. In general, when we add a quantity of acid equal to 100 times the weight of acid contained in the yeast, fermentation does not take place. It was necessary to add 200 equivalents of hydrocloric or tartaric acid to attain this result, although ten equivalents of these acids are sufficient to render the fermentation slow and incomplete.

Action of bases.—M. Dumas having set going at the same time eight experiments of fermentation, differing from each other only by increasing additions of ammonia from 0· (quantities equivalent to 0, 1, 2, 3, 4, 8, 16, 24 times the acid of the yeast employed), observed that in the first five vessels the fermentation was equal. It is only in proportions equal to eight, or even sixteen times the acid of the yeast, that it begins to proceed more slowly; but at the end of some hours the yeast has again become acid. In the flask containing a quantity of ammonia equivalent to twenty-four times the acidity of the yeast, no fermentation takes place.

Yeast seems to possess the power of producing or exhaling an acid which neutralizes the bases in contact with it; but this power is limited. In fact, by adding slaked lime or calcined magnesia in quantities equal to half the weight of the yeast there will be no fermentation. The oxides of zinc, of iron, or even litharge, have, on the contrary, no influence. The alkalies tend to arrest fermentation, but do not destroy it, except when the proportion is rather large. Sodium carbonate has no effect, unless we raise the proportion to 70 grammes for 10 grammes of yeast, and 200 cubic centimètres of sweetened water containing $\frac{1}{10}$ of sugar. Magnesium subcarbonate has no action.

Action of salts.—M. Dumas allowed yeast to remain three days in saturated solutions of various salts, and then tried its action on the solution of sugar-candy; his experiments led him to class salts in four categories, viz:—

1st. Those under whose influence the fermentation of sugar is entire, and more or less rapid—

Potassium sulphate	Sodium phosphate
" chloride	" sulphate
" phosphate	" bisulphate
" sulphovinate	" pyrophosphate
" sulphomethylate	" lactate
" hyposulphate	Ammonium phosphate
Calcium hyposulphate	Magnesium sulphate
Potassium formiate	Calcium chloride
" tartrate	" phosphate
" bitartrate	" sulphate
" sulphocyanide	Strontium chloride
" ferrocyanide	Alum
" ferricyanide	Zinc sulphate
	Cupric sulphate at $\frac{1}{10000}$

2nd. Those under whose influence there is partial fermentation of the sugar, more or less retarded—

Potassium bisulphite	Borax
" nitrate	White soap
" butyrate	Ammonium nitrate
" iodide	" tartrate
" arseniate	Seignette salt
Sodium sulphite	Barium chloride
" hyposulphite	Ferric sulphate $\frac{1}{350}$
Potassium hyposulphite	Magnesium sulphate $\frac{1}{350}$

3rd. Those under whose influence there is greater or less change of the sugar, without fermentation—

Potassium nitrate	Marine salt
„ chromate	Sodium acetate
„ bichromate	Sal ammoniac
Sodium nitrate	Mercuric cyanide.

4th. Those under whose influence there is neither change nor fermentation—

Potassium acetate	Sodium monosulphide.
„ cyanide	

Thus among the salts studied, there are some which favour fermentation, up to a certain point, or at least allow it to run its entire course without interruption; such as potassium tartrate. There are others which retard fermentation and render it incomplete, the action being arrested, when the liquor still contains much changed sugar in contact with it.

There are some which do not allow it to be set up, although the sugar has been partially changed.

There are also some, which not only do not allow fermentation to be set up, but which even oppose the change of the sugar.

Strychnine has no effect on the properties of yeast.

CHAPTER VII.

CAN NOTHING BUT ALCOHOLIC YEAST EXCITE ALCOHOLIC FERMENTATION?

WE must now enter on a question of great importance in the history of fermentation in general. Admitting with M. Pasteur, and all those who have preceded and followed him in this inquiry, that alcoholic fermentation, and the other phenomena of the same order, such as lactic, butyric, fermentations, &c., are palpable manifestations of certain physiological functions of ferments, or organisms of an inferior order, we may ask, whether the power of resolving glucose into alcohol and carbon dioxide, or of changing it into lactic acid, and that again into a mixture of hydrogen, carbon dioxide, and butyric acid, belongs, for each special fermentation, only to a single organism, to a single ferment, or, at least, to species very nearly allied, as we have seen in the species of the genus *Saccharomyces;* or whether these reactions are the result of cell life in general, when organic cells are placed under special conditions. On this hypothesis, ferments would have no advantage over other living cells, than that of showing these manifestations in a more energetic and intense degree.

It is easy to perceive, *à priori,* what new horizons would be opened for biological chemistry by such an idea, if it were correct, and what a vast field of experiments would be offered to chemical physiology.

There is no doubt that this thought must have presented itself involuntarily, as we may say, to the minds of men of science accustomed to reflect on these delicate questions of the reactions in living organisms; but M. Pasteur has the honour of first clearly expressing it, and supporting it by positive experiments. (Pasteur, Comp. Rend. de l'Acad. des Sciences, vol. 75, p. 784.) These experiments were undertaken in order to prove that the alcoholic fermentation of sugar may be excited by other organisms than the cells of *Saccharomyces*, and notably by the elementary cells of larger plants, such as are found in fruits, leaves, &c. The researches of M. Lechartier and M. Bellamy, on the alcoholic fermentation of fruits (Comp. Rend. de l'Acad., 1869, and 1872, vol. 75, p. 1,203; 1874, vol. 79, p. 949, and 1,006), have been directed to the same end, and lead, as we shall see, to this important consequence, that the elementary organs of plants in general are endowed, though in a less degree than the cells of yeast, with the property of exciting alcoholic fermentation.

Analogous facts had been observed by other experimentalists; thus, M. Bérard informed us, as early as 1821, that when fruits are placed in air, or in oxygen gas, a certain volume of this gas disappears, whilst at the same time there is a formation of a nearly equal volume of carbon dioxide. If these fruits are left, on the contrary, in carbon dioxide or in any other inert gas, there is still a formation of carbon dioxide in a notable quantity, "as if by a sort of fermentation."

M. Frémy had also shown that when we leave barley

in a solution of sugar in water, an incontestable intracellular fermentation takes place in the interior of the corn. (Comp. Rend. de l'Acad. des Sciences, vol. 75, p. 976 and 1,060.)

The same investigator observed the production of alcohol and carbon dioxide in the interior of fruits, such as pears and cherries; but guided by theoretical ideas of fermentation, which we will presently explain, he did not give to his interesting experiments the interpretation which they may receive.

M. Frémy admits that ferments are formed under the influence of the living organism. "Like all organisms in process of development," said he, "alcoholic ferment may present itself under the most various forms; it already exists, but in an imperceptible state, in the expressed juice of the grape, which appears to be transparent; it soon shows itself as very attenuated microscopical corpuscles; then undergoing a fresh development, it is precipitated at the bottom of the liquid, in the well-known form of grains of yeast. I have frequently examined with the microscope the juice and the parenchyma of fruits, before or after their fermentation, and I affirm that I have found an innumerable quantity of corpuscles which have all the appearance of organic ferments."

Thus we see, from this extract, that the intracellular fermentation of the fruit is not an immediate consequence of the biological functions of the living cell; there is an intermediate principle, the ferment, produced by the organism, which excites the decomposition.

He ought, in a historical point of view, to mention that the experiments of M. Lechartier, and M. Bellamy, preceded those of M. Pasteur; we will here mention the results obtained by these observers.

First, *M. Lechartier's Researches.* — This observer placed fruits (pears, apples, lemons, cherries, chestnuts, medlars, potatoes, grains of wheat, linseed, currants) in test-jars, connected with smaller test-tubes, placed over a mercurial trough. Under these conditions, the whole of the oxygen of the confined air in which the fruits are placed is absorbed. This absorption is accompanied and followed by a considerable production of carbon dioxide gas. The disengagement of carbon dioxide is generally divided into two distinct periods; in the first, after the absorption of the oxygen gas of the air which remained in the test-glasses, the disengagement of carbon dioxide proceeds, at first, in a regular and uniform manner, then it slackens, and stops for a certain time, to be afterwards renewed with an increasing rapidity, greater than that observed during the first period. This continues for several months. At this time, if we have taken the precaution to experiment on fruits, isolated from each other, and kept from contact with the sides of the vessel containing them; if, besides, we have been careful to prevent any deposit of liquid on the surface of the fruit, we may ascertain the production of notable quantities of alcohol, easily separated by distillation, after the fruit that has been experimented on has been crushed into pulp; and what is more, a careful microscopical examination of the parenchyma reveals no trace of alcoholic ferment.

Thus, on November 12, two pears, one weighing 157 grammes, and the other 125, were suspended separately, each in a glass jar, well corked, and provided with a tube by which gas might escape. Calcium chloride had been previously placed at the bottom of the jars, to maintain around the fruits an atmosphere unsaturated with the vapour of water.

The test-jars were opened on July 19: 1,762 cubic centimètres of gas, and 2·62 grammes of alcohol were collected. The pears had preserved their colour, their skin was wrinkled, but not damp. Their consistence and smell were like those of mellow pears. They had lost together 134 grammes of water, yet they still contained it in the proportion of 69 per cent. of their weight.

Microscopical observations, made at different distances from the centre, could discover no alcoholic ferment. The disengagement of carbon dioxide did not take place in a regular manner; from March 3 to April 8, there were only 28 cubic centimètres evolved; and from April 8 to July 19, there was absolutely none. The existence of alcoholic ferment within the pear is incompatible with all cessation of activity during so considerable an interval of time. This fact, therefore, corroborates the negative results obtained by means of microscopical examination.

In other experiments, arranged in the same manner, but continued much longer, M. Lechartier and M. Bellamy saw the disengagement of carbon dioxide resumed with increasing activity, after a shorter or longer period of cessation. In these cases, we can clearly ascertain the presence of ferment. We may suppose that the spores and germs of ferment, which exist on the surface of all fruits, as now seems incontestable, have been able to penetrate into the inside of the fruit by means of artificial fissures, and meeting there an appropriate medium, have budded and set up the alcoholic fermentation due to ferment.

The different fruits mentioned above all gave analogous results, and the experiment before described may serve as the type of the rest, which were very numerous, and were published by the authors; we will abstain from further allusion to them here.

The cessation of the disengagement of carbon dioxide may last a long time; thus, pears have been preserved inert, at the season of the year when the heat is greatest, during a time varying from 31 to 272 days; and when an end was put to the experiments, there was no reason to suppose that this state of things would be modified.

After this internal action, the fruits had undergone great modifications.

(1.) As soon as the fruits were exposed to contact with the air, they become brown throughout their whole mass, like fruits that have become "sleepy" or rotten. The leaves assume the colour and aspect of dead leaves.

(2.) The cellular tissue is either partly or completely disaggregated. Thus "duchesse" pears at the end of a year, were like a mass of syrup covered with a skin.

(3.) A fruit which has lost its activity does not regain it, even after being placed again in contact with the air.

(4.) The germ contained in the fruit participates in the decay, and the seed loses its property of germination.

It appears then, as these writers say, that at the

moment when fruits, seeds, and leaves are detached from the plant which bears them, life is not extinct in the cells of which they are composed. This life goes on, sheltered from the air, consuming sugar and producing alcohol and carbon dioxide. The moment when carbon dioxide ceases to be formed, is that at which all the vitality of their cells is destroyed. Fruits, seeds, and leaves may then remain for an indefinite period in an inert state, if no organic ferment is developed in the interior.

As to beet-root and potatoes, an especial fact has been observed which ought to be mentioned.

The whole of the phenomena are the same as with apples and pears; when the period of cessation arrives, no alcoholic ferment is observed, but in the acid liquid which impregnates the mass of their softened or disaggregated tissues, we find bacteria of different sizes, and yet we can detect no disengagement of carbon dioxide.

Secondly, *M. Pasteur's Experiments.* — The experiments made by M. Pasteur, while they verify the preceding conclusions, were especially intended to establish particular views respecting fermentations, and a general theory of these biological manifestations. The moment has now arrived to allude to them, and discuss them. We cannot do better than quote the exact words of the author (Pasteur, Comp. Rend. de l'Acad. des Sciences, vol. 75, p. 784, et seq.).

"I have been inclined for a long time to consider fermentations, properly so called, as chemical phenomena, co-relative with physiological actions of a peculiar nature. Not only have I shown that their ferments are not dead, albuminoid matter, but actual living organisms; I have also excited the fermentation of sugar, lactic acid, tartaric acid, glycerine, and, in more general terms, of all fermentable matters, *in media* exclusively mineral, an incontestable proof that the decomposition of fermentable matter is correlative with the life of the ferment, that it is one of its essential aliments; for instance, under the conditions to which I refer, it is impossible, that in the constitution of the ferments which arise, there can be a single atom of carbon which is not taken from the fermentable matter.

"That which distinguishes the chemical phenomena of fermentations from a number of others, and particularly from the functions of ordinary life, is the fact of the decomposition of a weight of fermentable matter much greater than that of the weight of ferment in action.

"I have long suspected that this peculiar character may be allied to that of nutrition without contact of free oxygen. Ferments must be living organisms, but of a peculiar nature, in this sense, that they have the property of exercising all the functions of their life, not excepting their multiplication,* without necessarily employing the oxygen of the atmospheric air.

"Let us call to mind those curious infusoria which cause butyric or tartaric fermentation, or certain kinds of putrefaction, and which not only are able to live and multiply without contact with oxygen, but which perish and cease to excite fermentation if we dissolve this gas in the medium in which they feed. This is not all. By careful experiments, made with the yeast of beer,

* Not that of sporulation. Note by M. SCHÜTZENBERGER.

I have shown, that if the life of this ferment were carried on partially by the influence of free oxygen gas, this little cellular plant lost, in proportion to the intensity of this influence, a part of its fermenting character, *that is to say, that the weight of yeast, which is produced under these conditions during the decomposition of sugar, increases progressively, and approaches the weight of decomposed sugar, in exact proportion as its life goes on in the presence of increasing quantities of free oxygen gas.*

"Guided by all these facts, I have been gradually led to look upon fermentation as a necessary consequence of the manifestation of life, when that life takes place without the direct combustion due to free oxygen.

"We may partially see, as a consequence of this theory, that every being, every organ, every cell, which lives, or continues its life, without making use of atmospheric air, or which uses it in a manner insufficient for the whole of the phenomena of its own nutrition, must possess the characteristics of a ferment with regard to the substance which is the source of its total or complementary heat. It appears that this substance must necessarily be oxygenated and carbonated, since, as I stated just now, it serves as food to the ferment. I have just brought to the support of this new theory, which I have already several times brought forward, though with some hesitation, since the year 1861, some new facts which will now, I hope, cause it to be received."

In a saccharine liquid, suited to the nourishment of ferments, contained in a vessel which allows an artificial sowing, while it prevents the spontaneous sowing of aerial germs, M. Pasteur placed on the surface a trace of pure *Mycoderma vini.** On the following days the growth covers by degrees all the liquid, under the form of a continuous film. We can easily ascertain that, under these conditions, the development of the *Mycoderma* gives rise to an absorption of atmospheric oxygen, which is replaced by an almost equal volume of carbon dioxide, and that, on the other hand, no alcohol at all is formed.

M. Pasteur having already pointed out that the *Mycoderma vini* has the property, when it grows on the surface of an alcoholic liquid, of burning the alcohol already formed, yielding not acetic acid or aldehyde, but water and carbon dioxide, it appears to us that we cannot conclude from the absence of observed alcohol in the experiment, that none has been formed. It is quite possible that it may have undergone a complete combustion at the very moment of its elimination. The volume of carbon dioxide disengaged in the experiments was nearly equal to the volume of oxygen consumed; therefore, unless we suppose that the combustion only acted on the hydrocarbons, or on carbon itself, we cannot explain this equality; it might be due to the disengagement of carbon dioxide under the influence of an alcoholic fermentation, one of the characteristics of which, alcohol, was burnt as fast as it was produced. However this may be, let us return to M. Pasteur's experiment. Let it be repeated exactly under the same

* In a publication previous to this, M. Pasteur had announced that the sedimentary yeast was produced by the *mycoderma vini*, deprived of contact with oxygen; M. Pasteur has reconsidered this opinion, which he no longer thinks thoroughly well founded.

conditions, with this difference only, that when the film is complete, the vessel be shaken, so as to break up this film, and sink it as far as possible, for the *fatty matter* which accompanies it hinders its being entirely wetted.

On the next day, and often after only a few hours, when the experiment is made at a temperature of from 25° to 30° C. (77° to 86° F.), we see small bubbles of gas continually rising from the bottom of the vessel, which show that fermentation has commenced in the saccharine liquid. It continues during the following days, although very feebly, and the presence of a sensible quantity of alcohol in the liquid is readily ascertained. A careful examination of the cells, or joints of the submerged *Mycoderma*, by means of the microscope, shows that these joints do not reproduce.

This latter fact seems to disagree with the opinion so many times advanced elsewhere by the writer, that alcoholic fermentation is always accompanied by the development and multiplication of ferment, and that it is a consequence of the formation of buds.

The experiment which we have just described in detail, quoting almost literally the words of the author, allows us to conclude with certainty that the *Mycoderma vini* may play the part of alcoholic ferment, when it is placed in contact with a saccharine medium ; but we cannot admit the other consequences that M. Pasteur tries to draw from it, as proofs of the theory before cited.

We may say the same thing of the experiments on fruits placed entirely, as M. Pasteur has done, in an atmosphere of carbon dioxide, or of those of M. Lechartier, in which this condition was rapidly obtained by the absorption of the oxygen of the surrounding air.

They prove irrefragably that vegetable cells can produce alcohol and carbon dioxide at the expense of sugar, and can act like the yeast of beer, but much less energetically; but nothing in the description given of their experiments by M. Pasteur and M. Lechartier can induce the reader to admit, that this production of alcohol, this cellular fermentation, only commences from the very moment when the cell is either partially or entirely removed from contact with oxygen.

Even admitting, which the writers quoted do not assert, that they have directly ascertained the absence of alcohol in the fruit, as long as it was not kept from the influence of oxygen, I still have the idea that the alcohol formed may have been burned, and may have disappeared under the form of water and carbon dioxide. Besides this the yeast of beer, the *Saccharomyces cerevisiæ* shows us in the most striking and incontestable manner, an instance of a cell which excites alcoholic fermentation, even in the presence of free oxygen dissolved in the medium; and now that it is no longer recognized as the sole producer of alcoholic fermentation, I do not see why we should wish to preserve for it so exceptional a character as that of being the only exciter of it in the presence of oxygen. As to this latter fact, we have only to quote M. Pasteur himself, in order to prove its reality (Bullet., Soc. Chimique, 1861, p. 621). "Yeast already formed may bud and develop itself in a saccharine and albuminous liquid in the entire absence of air or of oxygen. Little ferment is formed, in this case, and a comparatively large quantity of sugar disappears, sixty or eighty parts for one of the ferment formed. *Fermentation is very slow under these conditions.* If the experiment be made in contact with air, and over a large surface, *the fermentation is rapid.* For the same quantity of sugar that disappears, much more yeast is produced. This is developed under these conditions with energy, but its character as a ferment tends to disappear. We find, in fact, that for one part of ferment formed, there will be only from four to ten parts of sugar transformed. The part of ferment played by this yeast, nevertheless, remains, and shows itself *very energetic,* if we cause it to act on sugar when not under the influence of free oxygen gas.

M. Pasteur explains the fact of a tumultuous activity at the commencement of the fermentation, by the influence of the oxygen of the air which is dissolved in the liquids when the action commences.

There seems to be some contradiction between the conclusions and the facts. On one hand, it is ascertained that fermentation in the presence of free oxygen is very rapid and tumultuous; on the other hand, it is concluded that yeast loses its character as a ferment in the presence of oxygen. In reality, this contradiction disappears if we measure, as M. Pasteur has done, the power of the ferment, not by the quantity of sugar decomposed by the unit of weight of yeast, in the unit of time, which would lead us to consequences the reverse of those of this observer, but by the ratio between the weight of the sugar decomposed, and that of the ferment formed by budding. According to M. Pasteur, fermentation is a consequence of budding, and one of the phenomena may form the measure of the other.

Thus, by ascertaining that in fermentation without oxygen, the ratio between the sugar decomposed, and the yeast formed, is from 60 or 80 to 1, while in fermentation, in presence of oxygen, it is only 4 or 10 to 1, M. Pasteur thinks himself justified in concluding that, although in this latter case the fermentation is more active, the character as a ferment of the yeast has been lowered.

We may first remark that, in order to measure the energy of a ferment, M. Pasteur makes use, on this occasion, of an hypothesis which admits of discussion.

Is it not more natural to measure the energy of a ferment by the effect which it produces, and, in the present case, by the quantity of sugar decomposed in the unit of time by the unit of weight?

That which does come out clearly from the interesting facts observed by M. Pasteur, is, that yeast is able to multiply in a suitable albuminous saccharine medium, with or without the presence of free oxygen; the mutiplication is greatly favoured by oxygen, which explains why the ratio between the sugar split up and the yeast formed passes from 80.1, to 4.1, but at the same time the activity of the ferment is notably increased, and the fermentation is rapid and tumultuous.

The following experiment is also, in my opinion, opposed to M. Pasteur's theory. I have carefully studied, by means of a rapid and exact method of quantitative analyses of the dissolved oxygen, which has been described elsewhere, the respiratory phenomena of yeast. I have thus been able to ascertain that in a liquid medium containing oxygen in solution, the quantity of oxygen absorbed by one gramme of yeast, in the unit of time, and at the same temperature, was constant, whether the liquid be saturated or supersaturated with oxygen, or contain a quantity of oxygen less than is required for saturation. Thus, by diffusing one or two grammes of yeast in a litre of aërated water, we notice a sensible decrease of the respiratory power, only when the proportion of oxygen is less than one cubic centimètre per litre. On the other hand, these respiratory phenomena follow the same law in a saccharine and oxygenated medium containing nutritive albuminoid matter, but the intensity of the respiration is increased; that is to say, that the proportion of oxygen absorbed in the unit of time, by the unit of weight, is more considerable at the same temperature than in pure water. Finally, I have measured the alcohol produced, in a given time, in two fermentations placed identically under the same conditions, and with this difference only, that in the one oxygen was continually kept dissolved. The proportion of alcohol was found, in this case, to be sensibly greater, which agrees with Pasteur's experiments; during the whole duration of the fermentation, we have been able to detect a rapid absorption of oxygen, the intensity of which corresponded with the general results obtained in many experiments.

If, as M. Pasteur asserts, the decomposition of sugar were the result of a respiration of the cells of yeast at the expense of the combined oxygen, recruiting the free oxygen, it seems evident that fermentation ought not to take place, or at least ought to be sensibly lessened, in the presence of oxygen; but the reverse of this is the case. The respiratory power of yeast is independent of the quantity of oxygen contained in the medium in which it lives; it only varies with the temperature, and the more or less favourable conditions of nutrition, as well as with the more or less perfect state of health of the cell.

Yeast placed in a saccharine and oxygenated medium breathes as actively, or even more so, than in pure water; it therefore fully satisfies its respiratory power at the expense of free oxygen, and yet it excites the fermentation of the sugar as actively, or even more so, than in a medium which is not oxygenated. The conclusion to be drawn from this is plain; the respiratory power, and the fermenting power, are two qualities inherent in the cell of *Saccharomyces*, but independent of each other, in this sense, that one of these phenomena is not the consequence of the other. The energy of these two powers depends on conditions more or less favourable to the nutrition and development of the yeast; it is not, therefore, surprising to see them increase or grow less together. In other words, they are not, as M. Pasteur supposes, the two variable terms of a constant sum, of which the one vanishes when the other attains its maximum value; on the contrary, all the facts tend to prove that these two values grow weak, are destroyed, or attain their maxima at the same time, under the influence of the same causes. By weakening the vitality of the yeast, we diminish at the same time its respiratory power, and its power as a ferment; on the contrary, by placing the yeast under the best physical and chemical conditions of nutrition, we see the two factors increase side by side, and attain their maximum together. Let us, however, hasten to add, that these two phenomena are necessarily dependent on each other; and we cannot deny this without contradicting that which we have just said.

In fact, sugar is a pabulum necessary to the development of the cells of the *Saccharomyces;* its presence is, therefore, indispensable to the nutrition of these cells. To deprive the yeast of sugar, or which amounts to the same thing, to place it so that it is impossible for it to act as a ferment, is to diminish its vitality, and to place it in a state comparable to starvation; therefore it is not surprising to see its respiratory power diminish, all other things being equal. Reciprocally, to prevent yeast from breathing free oxygen, is to deprive it of one of its most important functions, and again to place it under pathological conditions; and its activity as a ferment will be lessened, as M. Pasteur himself observes.

The following experiments all support this view. We know, by the researches of M. Béchamp, that the yeast of beer placed under conditions of starvation, that is to say, kept from contact with a saccharine liquid, undergoes a great modification, which has nothing in common with putrefaction. It grows soft, and converts its protoplasm, and a part of its contents, into soluble principles, among which M. Béchamp distinguished leucine and tyrosine, a soluble albuminous substance coagulable by heat, a ferment producing change, a gummy substance, phosphates, and acetic acid; these phenomena are further accompanied by the production of alcohol and carbon dioxide, as well as by a disengagement of pure nitrogen. I have myself closely examined this singular reaction, and have been able to show, in the extract of softened yeast, the presence of xanthine, hypoxanthine, carnine, and guanine.

After these first researches, I thought it would be interesting to study the modifications which took place in the yeast when thus changed as to its respiratory functions. For this purpose I made use of the methods described elsewhere, when treating of the respiratory intensity of yeast. A known weight of yeast is diffused in a known volume of aërated water, the original oxymetrical degree of which is known; this degree is again measured after a quarter of an hour, or half an hour's respiration. The following facts were observed.

Four hundred grammes of fresh yeast, containing 28·94 per cent. of solid fixed matter, were diffused in distilled water, so as to form 2 litres (about 3½ pints). This homogeneous liquid was divided into ten equal parts, and introduced into bottles nearly filled. These bottles were left to themselves in a stove heated to 20° C. (68° F.).

I did not consider it to be necessary to employ in these experiments the creosote water recommended by M. Béchamp, as I was unacquainted with the nature of its influence on oxymetrical analyses or on the respiratory power. I therefore ran the risk of seeing some vegetable productions make their appearance, which were foreign to the yeast, but the small proportion of which, when compared with the yeast, could not disturb the results by any very large error.

At the end of three days without nourishment, the first and second vessels were emptied on a filter which had been previously weighed; that which remained on the filter, was, in this case, well washed with water at 30° C. (86° F.), until it was exhausted, then dried and weighed. The weight of the filter being deducted, there were found, for the 40 grammes of original yeast 5·37 grammes, or 13·4 per cent. of solid fixed matter. The contents of the second bottle were boiled in water before filtration; the fixed residuum found was, 5·73 grammes, or 14 per cent.

After digestion for six days, followed by boiling, the third bottle gave a dry fixed residue of 4·16 grammes, or 10·4 per cent.

After a longer time, the diminution in weight of the solid insoluble materials of the yeast grew less by degrees, and became insensible. Fresh yeast, washed with lukewarm water, loses about 9·5 grammes per cent. of soluble matter. Thus, in five days, the yeast which had been kept without nourishment, transformed into soluble products, 18·5—9·5=9· grammes of insoluble matter, which resisted the first washing.

When it has reached this limit, and has been well washed with warm water till it runs off pure, yeast shows itself completely inactive with regard to oxygen. Thus fresh yeast, at 20° C. (68° F.), respired, per gramme and per hour, 1·4 cubic centimètre, whilst after exhaustion, by washing the absorption of oxygen, under the same conditions, did not exceed ·23 cubic centimètre. Yeast digested, without nourishment, for forty-eight hours and well washed, being placed in a pure saccharine medium, only excited a very slow and scarcely perceptible fermentation, as M. Béchamp had before observed.

Thus, digestion without nourishment, when sufficiently prolonged, and followed by thorough washing, affords us a very simple method of removing, almost entirely and simultaneously, the two most characteristic biological manifestations of the yeast-cell. Yet its vitality is not destroyed, if the experiment has not been carried too far.

If we add to aërated water, containing yeast which has been digested and washed, and which only acts on oxygen in an almost imperceptible manner, some water with which digested yeast has been washed, we shall see the respiratory phenomenon, at first very feeble, grow more and more decided, and resume an activity equal, and even superior, to that of fresh yeast simply diffused in pure water. The fresh yeast itself respires more energetically if, instead of pure water, we employ aërated water containing soluble products, derived from the digestion of yeast.

On the other hand, if we add the soluble products of yeast to the mixture of digested and washed yeast and saccharine water, a mixture which gives rise only to an insensible fermentation, we see, in a very short time, the decomposition of the sugar manifest itself by a more abundant disengagement of carbon dioxide.

It follows, from the whole of the facts noticed in this chapter, that we can no longer regard alcoholic fermentation as the simple result of the biological intervention of a special living organism, characterized by its form and by the conditions of its development, which alone has the power to split up some one particular organic compound in some one particular way.

The various kinds of fermentation appear to us, more and more, as particular cases of the chemical activity of living cells, and the ideas enunciated at the beginning of this work on fermentation find a complete confirmation in the new facts studied in this chapter.

M. A. Fitz (Soc. Chim. de Berlin, January and February, 1873), has published a long memoir on alcoholic fermentation produced by the *Mucor racemosus.*

The ratio of the alcohol to the carbon dioxide was found, as the mean of many determinations, to be equal to $\frac{100}{123 \cdot 1}$, while in the alcoholic fermentation set up by yeast of beer, it is $\frac{100}{96 \cdot 3}$.

Distilled alcohol contains a little aldehyde; the residuum contains succinic acid; while the presence of glycerine is doubtful. Dextrin, inulin, and sugar of milk do not ferment, alcoholically, in the presence of *Mucor racemosus.*

With altered sugar, the fermentation set up by *Mucor racemosus* stops after a short time, so that half of the sugar remains intact.

The cause of this rapid check in the process of fermentation is merely the sensitiveness of this ferment to the influence of alcohol.

M. Fitz has, in fact, shown that fermentation ceases when the proportion of alcohol in the liquid reaches 3·5 to 4 per cent. by weight. In all other respects the *Mucor racemosus* behaves like yeast; it alters sugar, assimilates the nitrogen of ammoniacal salts, and reproduces at the expense of sugar (1·3 per cent. of the sugar decomposed).

M. Traube (Soc. Chim. de Berlin, vol. vii. p. 887, 1874) has studied the action of media containing no free oxygen on alcoholic ferments. His experiments lead to the following conclusions:—

The cell of the ferment is not developed in the absence of free oxygen, even in its most favourable medium, the must of grapes.

The ferment, in process of development, continues to increase in suitable media, even in the absence of all trace of oxygen, as M. Pasteur had already shown. The contrary assertions of Brefeld are erroneous.

M. Pasteur's theory, according to which the yeast, in the absence of air, takes from the sugar the oxygen necessary for its development, is not well founded; in fact, this development stops long before the greater part of the sugar is decomposed. Is it from the albuminoid matter that the ferment takes this oxygen in the absence of air?

Yeast sets up alcoholic fermentation in a solution of pure sugar in the absence of all trace of oxygen, but without developing. This is contrary to the affirmation of M. Pasteur that fermentation is bound up with the organization of the yeast, or is a phenomenon correlative to the vital activity of the cells.

CHAPTER VIII.

VISCOUS OR MANNITIC FERMENTATION OF SUGAR.

LIKE all the phenomena which are spontaneously produced under the ordinary conditions of humanlife, viscous fermentation has been known for a long time, for it is developed in certain wines and natural saccharine juices, as the juice of beetroot, carrots, and onions, as well as in certain juleps and potions containing sugar and nitrogenous matter, and it becomes apparent by the viscosity of the liquid.

This reaction was studied by Braconnot (Ann. de Chim. (1), lxxxvi. p. 97, 1813), Desfosse (Journ. de Pharm., xv. p. 604), François (Ann. de Chim. et de Phys. (2), vol. xlvi. p. 212, 1829), Pelouze and Jules Gay-Lussac (Ann. de Chim. et de Phys. (2), vol. lii. p. 410, 1833), Kircher (Annalen. Ch. Pharm., xxxi. p. 337), Tilley and Maclagan (Philos. Mag., xxviii. p. 12), Boutron-Charlard and Frémy (Comp. Rend., xii. p. 708, 1841), and also by Pasteur (Bullet., Soc. Chim., p. 30, Paris, 1861).

Viscous fermentatation is excited, according to Pasteur, by a special ferment acting on glucose or cane-sugar, previously altered, transforming them into a kind of gum or dextrin, mannite, and carbon dioxide. Gum, mannite, and carbon dioxide are the only constant products of this reaction; the lactic acid, butyric acid, and hydrogen, which we see appear at the same time, are the products of other fermentations due to foreign ferments.

M. Peligot (Traité de Chimie de Dumas, vol. vi. p. 335, 1843) was the first to notice a special ferment, capable of producing viscous fermentation in saccharine solutions to which it is added. Pasteur observed, afterwards, that this ferment is formed of small globules united as in a necklace, whose diameter varies from ·0012 millimètres (·000047 in.) to ·0014 (·000055 in.) (fig. 20). These globules, sown in a saccharine liquid contain-

FIG. 20.—Viscous Ferment of Wine.

ing nutritive nitrogenous matter and mineral substances, always give rise to viscous fermentation: 100 parts of sugar give about 51·09 of mannite and 45·5 of gum; besides these, carbon dioxide is disengaged. These results may be shown by the following equation:—

$$\text{No. 1. } \underset{\text{Cane-Sugar.}}{25\,(C^{12}H^{22}O^{11})} + \underset{\text{Water.}}{25\,(H^2O)} = \underset{\text{Gum.}}{12\,(C^{12}H^{10}O^{10})} + \underset{\text{Mannite.}}{24\,(C^6H^{14}O^6)} + \underset{\text{Carb. diox.}}{12\,(CO^2)} + \underset{\text{Water.}}{12\,(H^2O)}$$

This gives, for 100 parts of sugar,—

Mannite	51·09
Gum	45·48
Carbon dioxide	6·18

M. Monoyer (Thèse pour le Doctorat en Medicine, Strasbourg, 1862) proposes to represent the production of mannite and of gum by two separate equations, applying to two distinct phenomena, independent of each other.

$$\text{No. 2. } 13\,(C^{12}\,H^{24}\,O^{12})^{*} + 12\,H^{2}\,O = 24\,(C^{6}\,H^{14}\,O^{6}) + 12\,CO^{2}$$

$$\text{No. 3. } 12\,(C^{12}\,H^{24}\,O^{12})^{*} = 12\,(C^{12}\,H^{20}\,O^{10}) + 24\,H^{2}\,O\,;$$

which, added together, reproduce Equation No. 1 (plus 12 molecules of water). The respective proportions of gum and mannite, corresponding to the preceding results, are obtained generally and with constancy when we sow the ferment before described.

In certain viscous fermentations, however, the proportion of gum is greater than that of mannite. In this case, according to M. Pasteur, the presence in the liquid of larger globules of a different kind is perceived. The same writer adds, that it may be possible that this second ferment transforms the sugar into gum only, without there being any fermentation of mannite. To prove the accuracy of this view, it would be necessary to be able to isolate the second ferment from the first, and to cause it to act separately: hitherto this separation has not been effected. However this may be, this variability in the proportion of the gum tells in favour of M. Monoyer's opinion.

The liquids which are most apt to produce viscous fermentation † can also undergo lactic and butyric fer-

* Perhaps better written, 13 $(C^{6}\,H^{12}\,O^{6} + C^{6}\,H^{12}\,O^{6})$ and 12 $(C^{6}\,H^{12}\,O^{6} + C^{6}\,H^{12}\,O^{6})$ as these formulæ represent the "changed" or "altered" cane-sugar, *i.e.*, cane-sugar that has taken up one molecule of water and split up into a mixture of dextrose and levulose.

† Decoction of the yeast of beer filtered, and with the addition of sugar;

mentation ; but, in this case, the organized forms of life which are developed in the liquid are of a different nature.

The gum which is formed under these conditions is more allied, by its characters, to dextrine than to gum arabic. Nitric acid converts it into oxalic acid without the production of mucic acid. The conditions of action necessary to these gummy and mannitic ferments are the same as those which suit alcoholic ferment. The most favourable temperature is 30° C. (86° F.).

water of flour, barley, and rice with sugar added. White wines are more subject to this change, called ropiness, than red wines : as this difference appears to be connected with the absence of tannin in the former, M. François proposes the addition of tannin to white wine, as a remedy for the ropy affection.

CHAPTER IX.

LACTIC FERMENTATION.

THE lengthened discussion into which we have entered with respect to the history of alcoholic fermentation allows us to be much more brief when speaking of the other phenomena of this order, which have been generally attributed, since the time of Pasteur, to the intervention of lower organisms. Except the chemical reaction, and the form of the ferment, the general characters of these fermentations are the same, for the very simple reason that the conditions of nutrition and of development of organic ferments do not vary within very wide limits.

By lactic fermentation is understood the transformation of certain sugars, as sugar of milk and grape-sugar into a sirupy acid, soluble in water, (lactic acid). By comparing the formulæ of the sugar which ferments, and of the lactic acid, the only product of its fermentation, we see that we have here a simple molecular transformation, or rather the splitting up of one molecule into two more simple equivalent ones. We have, in fact,—

$$\underset{\text{Sugar.}}{C^6\,H^{12}\,O^6} = \underset{\text{Lactic acid.}}{2\,C^3\,H^6\,O^3}$$

This transformation has been observed, long since, under many circumstances.

M. Boutron and M. Frémy were the first to consider it the result of a special fermentation, independent of the viscous fermentation with which it was confounded. (Ann. de Chim. et de Phys. (3), vol. ii. p. 257, 271.)

When milk turns sour spontaneously, the sugar of milk which it contains is converted into lactic acid, and it was from sour skimmed milk that Scheele first extracted lactic acid, as early as 1780. Braconnot found the same acid in rice left under water in fermentation, in the juice of beetroot, which, after having passed through the viscous fermentation, and a slight alcoholic fermentation, becomes sour, and produces lactic acid and mannite; he also found it in the water of fermentation of peas and boiled French beans, and in the sour water of baker's yeast. The same product is found in the sour water of starch manufacturers, and in sauerkraut. M. Frémy and M. Boutron, as well as M. Pelouze and M. Gélis, ascertained the best conditions for effecting the rapid transformation of sugar into lactic acid. According to their researches, lactic fermentation requires the presence of nitrogenous albuminoid matter in process of decomposition, and can only continue if the degree of acidity of the liquor is kept from exceeding certain limits. The end is best attained if we saturate it from time to time with sodium carbonate, or add previously a quantity of chalk, sufficient to neutralize all the acid which can be formed at the expense of the sugar.

We will now give, in a few words, the best process, as published by the writers who have studied this subject.

Three or four litres of milk are taken, to which from 200 to 300 grammes of sugar of milk is added, the liquor is left exposed to the air, in an open vessel, for several days, at a temperature between 15° and 20° C. (50°—68° F.). After this time the liquor is found to be acid; it is then saturated by sodium carbonate, and these saturations are carried on successively, at intervals of twenty-four hours, till all the sugar of milk is transformed. It is now boiled, filtered, and cautiously evaporated to a sirup. The remainder is treated with alcohol at 38°, which dissolves the sodium lactate. This solution is precipitated by a suitable quantity of sulphuric acid, which precipitates the soda; it is then filtered and evaporated. The remainder, saturated with lime, will give botryoidal crystals of pure calcium lactate. (Boutron and Frémy.)

Bensch's process is also to be recommended. It consists in dissolving three kilogrammes of cane-sugar (about 6·6 lbs. av.) and 15 grammes (·083lbs. av.) of tartaric acid in 13 kilogrammes of boiling water. The solution is then left to itself for some days; after which 60 grammes (·132lbs. av.) of old decayed cheese, diffused in 4 kilogrammes, (8·8lbs. av.) of curdled and skimmed milk, 1½ kilogrammes of washed chalk (3·3lbs. av.) are added. The whole is put in a warm place 30°—35° C. (86°—95° F.), and stirred from time to time. After eight or ten days, the liquid forms into a mass of calcium lactate. Ten kilogrammes of boiling water (22lbs. av.) and 15 grammes (·033lbs. av.) of slaked lime are then added; it is boiled, and passed through a cloth; then evaporated and crystallized. The crystals are pressed and purified by new crystallizations. The calcium lactate is then dissolved in water precipitated by a suitable quantity of sulphuric acid (210 grammes for 1 kilogramme of lactate) ; the calcium sulphate is separated, and the liquid is neutralized by zinc carbonate ; we thus obtain zinc lactate, which can be easily purified by crystallization, and is readily converted by sulphuretted hydrogen into lactic acid and insoluble zinc sulphide. Nine kilogrammes of sugar thus yield easily 10½ kilogrammes of calcium lactate.

The juice of fermented beetroot (Pelouze and J. Gay-Lussac), and the water in which sauerkraut has been washed (Liebig), will also serve for the extraction of lactic acid, by analogous methods, and by the use of chalk, or zinc carbonate. The terms of the reaction, and the conditions favourable to its production, were, therefore, well known ; but before the researches of Pasteur, only vague and ill-founded ideas were held respecting the cause which produced this phenomenon. The apparent absence of an organized ferment, which had not been found,* gave a powerful support to Liebig's theory.

Casein or albuminous matter seemed to act upon sugar only by its own decomposition, and it might be supposed that the molecular movement, the consequence

* The true lactic ferment had, however, been already surmised by Dr. Remak, (Remak Canstatt's Jahersbericht, i. 1841), and partly described by Blondeau (1848), "Journ. de Pharm." (3), vol. xii, 244 and 336. Remak had observed that the globules of beer-yeast are often mixed with smaller ones, which give rise to a special fermentation, and can, in no case, be transformed into globules of the yeast of beer. Blondeau points out in beer-yeast two kinds of cells, the ordinary ones of ·01 millimètre in diameter, setting up alcoholic fermentation, and cells about four times smaller, which give rise to lactic fermentation. Blondeau attributes to them both butyric fermentation and the ammoniacal fermentation of urea ; but in this he was mistaken.

of this spontaneous decomposition, was capable of transmitting itself to the saccharine principles contained in the liquor.

M. Pasteur, as early as 1857, guided by well-founded ideas as to the causes of alcoholic fermentation, and fermentation in general, sought for the lactic ferment; that is to say, for the organism which transforms sugar into lactic acid by the effect of its nutrition and development, and succeeded in finding and studying it, although it did not show itself so distinctly as alcoholic ferment.

"If we examine," said he (Ann. de Chim. et de Phys. (3), vol. lii. p. 407, Mémoire sur la Fermentation appelée Lactique), "with attention, ordinary lactic fermentation, we find cases in which we can recognize upon the deposit of chalk and nitrogeneous matter, spots of a grey substance sometimes forming a zone on the surface of the deposit. This substance is sometimes found sticking to the upper parts of the vessel, whither it has been carried by the movement of the gas. Its examination with the microscope scarcely enables us to distinguish it, unless we are prepared for it beforehand, from caseum, disaggregated gluten, &c.; so that nothing shows that it is a special substance, nor that it has been formed during the fermentation. Its apparent weight is always very trifling compared with that of the nitrogeneous matter primarily necessary for the production of the phenomenon. Indeed, it is so very often mixed up with the mass of caseum and chalk, that we should have no reason to suspect its existence. It is this, nevertheless, which plays the most important part."

To prove this, M. Pasteur thoroughly washed fresh

yeast with fifteen or twenty times its weight of boiling water. To this liquid, filtered with care, and containing the most appropriate nitrogeneous and mineral aliment for the nutrition of organic ferments, was added from 50 to 100 grammes of sugar per litre, and chalk. A trace of the grey matter mentioned above was sowed in it, carbon dioxide introduced instead of air, and the whole was kept at 30° or 35° C. (86° to 95° F.). In a short time, all the symptoms of lactic fermentation were developed, accompanied by those of a weak butyric fermentation.

When the chalk has disappeared, the concentrated liquid furnishes an abundant crystallization of calcium lactate, while the mother liquor retains calcium butyrate. Sometimes the liquor becomes very viscid. In a word, we have before our eyes a most characteristic example of lactic fermentation, with all the accidents and the usual complication of this phenomenon. At the same time that the reactions that have been described take place, the liquid, limpid at first, becomes troubled; a deposit appears which increases continually, as fast as the chalk is dissolved. We may substitute

FIG. 21.—Lactic Ferment.

in this experiment any plastic nitrogeneous substance, either fresh or decomposing, for the decoction of yeast.

The substance which is deposited, and whose production is correlative to the reactions known by the name of lactic fermentation, exactly resembles, when taken in mass, ordinary yeast, which has been drained of water and pressed. It is rather viscid, and of a grey colour. Under the microscope, it appears as if formed of small globules or short joints, either isolated or in a mass, forming flakes resembling those of certain amorphous precipitates (Fig. 21). The globules, much smaller than those of yeast of beer, are violently agitated by the Brownian movement, when they are isolated. When washed with much water by decantation, and then diffused in pure solution of sugar, it begins to render it acid at once, but progressively, and very slowly, because acidity greatly interferes with its action. If we add chalk, so as to maintain the neutrality of the medium, the transformation of the sugar is sensibly accelerated; at the end of an hour, gas begins to be disengaged; and the liquor becomes charged with calcium lactate and butyrate in variable quantities.

When the solution of sugar contains nitrogeneous matter, and salts suited to the nutrition of ferments, the lactic ferment is developed, and we obtain quantities which have no limit except the weight of the sugar and the albuminous substances employed. But little of this ferment is necessary to transform a considerable weight of sugar. Its activity is only weakened when we dry it, or boil it in water. This fermentation ought to be set up without access of air, in order that we may not be annoyed by vegetation and infusoria from without.

Lactic fermentation, under the conditions indicated by M. Pasteur, *when the lactic ferment is developed spontaneously*, is often more rapid than alcoholic fermentation, and shorter than in the ordinary process. (Boutron and Frémy, Bensch, &c.)

The most favourable temperature seems to be about 35° C. (95° F.).

If, without making any change in the conditions of the saccharine medium, which when sowed with the lactic ferment, produces lactic fermentation, we sow this medium with globules of yeast, we observe the appearance of alcoholic fermentation and the development of alcoholic ferment.

Finally, to complete the analogy between the two kinds of fermentation, M. Pasteur showed that lactic ferment can, like alcoholic ferment, be produced as if spontaneously in saccharine liquids of a composition adapted to its development. Thus in alcoholic fermentations in an artificial medium, sowed with traces of alcoholic ferment, M. Pasteur almost always saw lactic ferment and fermentation appear side by side with alcoholic ferment and fermentation.

In these cases of spontaneous production of lactic, in the presence of alcoholic ferment, the predominance of one or the other organism, and consequently of its effects, depends on the composition of the medium, being more or less suited to one or the other ferment, and notably so, on the neutral state of the liquid.

Thus, if we add magnesia to a mixture of solution of sugar and beer-yeast, there will be at the same time alcoholic and lactic fermentation with precipitation of magnesium lactate, and we shall see, mixed with the globules of yeast of beer, a considerable quantity of the small globules of lactic ferment. A medium, slightly alkaline, is more suited to the development of the new ferment; it is the same with respect to alcoholic ferment.

The concomitant development of alcoholic and lactic ferment explains why frequently the presence of lactic acid has often been observed as one of the products of the decomposition of sugar under the influence of beer-yeast.

Most careful and frequent experiments proved to M. Pasteur, that not even the smallest quantity of lactic acid is formed in normal alcoholic fermentation.

When it does appear, which is very rarely, unless we choose favourable conditions, we may be sure that the beer-yeast is mixed with lactic ferment, which can easily be recognized by its form, its size, and its movement.

To be certain of the presence or absence of lactic acid, we operate as has been before explained (page 30) when treating of succinic acid and glycerine. The ethereal alcoholic solution of the extract of the fermented liquid is evaporated; the remainder is saturated with lime-water; the liquid is again evaporated, and the remainder is taken up by ethereal alcohol. The residuum is finally heated with boiling alcohol at 95 per cent., which dissolves the calcium lactate.

The following substances are susceptible of undergoing lactic fermentation, the glucoses and those bodies which are convertible into glucoses. The saccharoses which most easily undergo alcoholic fermentation, are those which are with the greatest difficulty transformed into lactic acid, and *vice versâ*. Sugar of milk does not readily produce alcohol, while it is much more easily resolved into lactic acid, although according to Lubolt (Journ. J. Prakt. Chim. vol. 77, p. 282) and Proust (Ann. de Chim. et Phys. (2), vol. 10, 29) the action is very slow, sugar of milk being found in the liquor up to the end of the fermentation.

Sorbin, inosite, mannite, and dulcite do not undergo alcoholic fermentation, but are affected by lactic ferment.

Calcium malate, and, in general, all substances whose fermentation produces butyric acid, seem to have the power of yielding lactic acid.

In the fermentation of calcium malate, carbon dioxide is disengaged.

$$\underset{\text{Malic Acid.}}{C^4\,H^6\,O^5} = \underset{\text{Lactic Acid.}}{C^3\,H^6\,O^3} + \underset{\text{Carbon dioxide.}}{C\,O^2}.$$

CHAPTER X.

AMMONIACAL FERMENTATION.

UREA ($C H^4 N^2 O$) is, as is well known, one of the most important constituents of urine. It appears in it as the principal form under which nitrogen is eliminated from the animal organism.

This body, of simple composition, crystallizable in long prisms, only differs from ammonium carbonate by the elements of water.

We have, in fact,—

$$\underset{\text{Urea.}}{(CH^4\ N^2\ O)} + \underset{\text{Water.}}{H^2O} = \underset{\text{Carbon dioxide.}}{CO^2} + \underset{\text{Ammonia.}}{2\ N\ H^3}.*$$

By boiling in alkaline, or even in pure water, this phenomenon of hydratation is produced more or less quickly. We even observe it with the more complex compounds, known under the name of ureids, and in which we must admit that the molecule of urea is associated with other organic groups, such as uric acid, alloxane, creatine, &c.

A solution of pure urea in water may be preserved for a long time without change. It is not so with the natural solutions of urea, such as urine, which contains

* Or better, $\underset{\text{Urea.}}{CH^4\ N^2\ O} + \underset{\text{Water.}}{2\ H^2\ O} = \underset{\text{Ammonium carbona.}}{(NH^4)^2\ C\ O^3}$.

salts and other nitrogeneous principles more similar to albuminous substances.

It is well known that, after a longer or shorter time, according to the conditions of temperature, and the state of health of the individual who has excreted the urine, this liquid, after its emission, becomes alkaline, instead of acid as it was before; at the same time, it exhales a very decided odour of ammonia; at this moment the urea has disappeared, or is on the point of disappearing entirely; we find in its place an equivalent quality of ammonium carbonate.

In certain cases, the urine is already alkaline and ammoniacal, when in the bladder.

The name of ammoniacal fermentation has been given to this spontaneous transformation. (Dumas, Traité de Chimie, vol. 6, p. 380.)

According to the observations of Müller (1860, Journ. für Prakt. Chem., 81, p. 467) and of Pasteur (Comp. Rend., vol. 50, p. 869, May, 1860), the transformation of urea into ammonia and carbon dioxide is due to the intervention of a special organic ferment, produced by one of the *torulacei* and formed of chaplets of globules very similar to those of beer-yeast, but much smaller; their diameter is about $\frac{1}{5000}$ of a millimètre (,0000078 in.). M. Van Tieghem has very thoroughly studied this ferment. It is found in the white deposit left at the bottom of urinals. M. Jaquemart (Ann. Chim. Phys. (3), vol. 7, p. 149, 1843) had already noticed that this deposit was very apt to excite this transformation.

Long study of the organic products which are developed in urine exposed to air, has convinced M. Van Tieghem of the constant presence of a (torulaceous) ferment, whenever urea ferments; and of the intimate connection which exists between its easy or difficult development, and the rapid or slow transformation of the urea.

"In the case, seldom realized, in which this torulaceous growth is developed alone, the liquid remains limpid, the fermentation is rapid, and the deposit which forms at the bottom of the vessel is composed exclusively of chaplets and masses of globules mixed with crystals of urates and of ammonium-magnesium phosphate. If the torulaceous growth is only accompanied by infusoria, as is usually the case, the fermentation, though somewhat slower, is still easy; but if there appear, besides the infusoria, vegetable productions in the liquid and on the surface, the torulaceous growth is developed with difficulty, and the transformation is very slow; the liquid may remain acid or neutral for months together.

"If, instead of leaving the urine to the variable chances which the sequence of the presence of the germs in the air involve, we place it in a stove in a corked bottle, having added to it a trace of the deposit of a good fermentation, all accidental variations disappear, and the phenomenon takes place always in the same manner; one or two days are sufficient for the urea to disappear, and at the same time the torulaceous growth alone is developed.

"The transformation of urea in urine is therefore correlative to the life and development of an organic vegetable ferment. This ferment is developed within the liquid itself, and especially at the bottom of the vessel, where, by its accumulation, it forms a whitish deposit, and is composed of chaplets, or small masses of spherical globules, without granulations, without any distinct envelope of the contents, and which appear to develop by budding; their diameter is about ·0015 millimètres (·000059 in.)."

M. Van Teighem considers that he has proved by direct experiment, that the splitting up of hippuric acid by hydratation, into benzoic acid and glycocol, which is observed in the urine of the herbivora after emission, is due to a fermentation analogous to that which splits up the urea. The active ferment must be identical with the ammoniacal ferment. Thus, ammonium hippurate dissolved, either in yeast-water or in solution of sugar containing phosphates, is always split up in consequence of the development of a microscopic vegetable organism identical with the torulaceous growth described above. (Comp. Rend. de l'Acad. des Sci., vol. 58, p. 533.)

According to Müller, the activity of the phenomenon is proportionate to the number of globules. When to a mixture of sugar and urea, dissolved in water, we add beer-yeast, we always see the small globules of ammoniacal ferment make their appearance, as soon as the liquid shows an alkaline reaction (Pasteur). The yeast of beer, by itself, decomposes the sugar without exciting the decomposition of the urea. The most suitable temperature is that of the human body, 37° C. (99 F°).

Finally, as is the case with the must of grapes and the wort of beer, the ferment does not pre-exist in the urine; it must be brought from without, in the form of germs.

The necessity for ammoniacal ferment, in order to transform the urea found in urine, under ordinary conditions of temperature, raises a medical question of great importance. In fact, many cases of disease have been observed in which the urine was more or less ammoniacal in the bladder. This change usually accompanies either vesical or renal injuries, or serious general diseases, such as typhoid fever. When the urine is ammoniacal after the catheter has been introduced, it is to be feared that the instrument has served as a vehicle, and carried into the bladder the germs of the torulaceous growth, thus setting up infection. Ammoniacal urine, from a patient at la Charité, was examined by M. Gayon, at the very instant of emission ; he found in it innumerable organisms : this examination took place some days after the patient had been relieved by the catheter. However, in the opinion of surgeons, the urine often shows itself to be ammoniacal without any previous operation of this kind ; for example, in cases of retention, although the long continuance of the liquid in the bladder may not be the only condition of this phenomenon. In fact, M. Verneuil did not find the urine ammoniacal in the case of an hysterical young girl, in whose case the catheter had been passed, after several days' retention.

As M. Pasteur has observed, the canal of the urethra, with relation to the infinitely small germs which act as ferments, may be compared to the Thames tunnel, through which they can pass and repass easily. If then, we should feel astonished at anything, it is rather at finding the ammoniacal infection occur so infrequently. It is, perhaps, because the urine is usually acid, and that this acid is unfavourable to the development of the germs of fermentation.

However this may be, M. Pasteur, without being

able to give any positive proof, is inclined to believe that the decomposition of urea is not possible, without the presence of a peculiar ferment, and that there is no purely chemical reaction in the human body, capable of giving rise to ammonium carbonate in the urine. This conviction has been the result of long study of ferments, and of his great experience. He owns, however, that the definite solution has not yet been obtained. (*See* on this subject the discussion on ammoniacal urine at the Académie de Médicine, Moniteur Scientifique (3), vol. 9, p. 143.)

In conclusion, the transformation of the urea in urine, at the ordinary temperature, may certainly be caused by a special ferment. It is possible, but as yet not certainly proved, that the presence of this ferment may be indispensable.

CHAPTER XI.

BUTYRIC FERMENTATION AND PUTREFACTION.

A GREAT number of chemical compounds are susceptible of fermenting butyrically, that is to say, yielding butyric acid as a product of their transformation, when they are placed under suitable conditions.

Such are lactic acid, and all substances capable of undergoing lactic fermentation—sugars, amylaceous matter, tartaric, citric, malic, mucic acids, and albuminoid substances.

It is easy to represent by equations the production of butyric acid at the expense of the greater part of these bodies, which are well defined and of well-known composition.

Thus sugar and lactic acid would give—

$$\underset{\text{Glucose.}}{C^6 H^{12} O^6} = \underset{\text{Lactic acid.}}{2\ C^3 H^6 O^3} = \underset{\text{Butyric acid.}}{C^4 H^8 O^2} + \underset{\text{Carb. dioxide.}}{2\ CO^2} + \underset{\text{Hydrogen.}}{H^4}$$

For malic acid we shall have—

$$\underset{\text{Malic acid.}}{2\ C^4 H^6 O^5} = \underset{\text{Lactic acid.}}{2\ C^3 H^6 O^3} + \underset{\text{Carbon dioxide.}}{2\ CO^2} = \underset{\text{Butyric acid.}}{C^4 H^8 O^2} + \underset{\text{Carb. diox.}}{4\ CO^2} + \underset{\text{Hydrogen.}}{H^4}$$

Tartaric acid would be resolved into butyric acid, carbon dioxide, and water:

$$2\, C^4\, H^6\, O^6 = C^4\, H^8\, O^2 + 4\, CO^2 + 2\, H^2\, O$$

Tartaric acid. Butyric acid. Carbon dioxide. Water.

Perhaps lactic acid is formed first:

$$3\, C^4\, H^6\, O^6 = 2\, C^3\, H^6\, O^3 + 6\, CO^2 + H^6$$

Tartaric acid. Lactic acid. Carb. diox. Hydrogen.

According to M. Personne, citric acid would be transformed previously into lactic and acetic acid:

$$4\, C^6\, H^8\, O^7 + 2\, H^2\, O = 3\, C^2\, H^4\, O^2 + 4\, C^3\, H^6\, O^3 + 6\, CO^2$$

Citric acid. Water. Acetic acid. Lactic acid. Carb. diox.

Mucic acid, which differs from citric acid only by containing more water, is resolved, like it, into acetic and butyric acids, carbon dioxide, and hydrogen:

$$3\, C^6\, H^{10}\, O^8 = 3\, C^2\, H^4\, O^2 + C^4\, H^8\, O^2 + 8\, CO^2 + H^{10}$$

Mucic acid. Acetic acid. Butyric acid. Carb. diox. Hydrogen.

In other analogous phenomena, which go on at the expense of the same substances, we find acids homologous to butyric acid; in the same way as, in alcoholic fermentation, alcohols are formed with higher equivalents than that of ethyl-alcohol.

Thus, under the conditions of lactic fermentation, we see the formation of propionic acid, acetic acid, and valerianic acid at the expense of sugar or of starch.

Glycerine, placed in contact for a long time with beer-yeast, also furnishes propionic acid, mixed with formic and acetic acids. M. Monoyer represents in a general manner the production of fatty acids at the expense of sugar by the equation—

$$\frac{n - 1}{3}\,(C^6\, H^{12}\, 6^6) = [C^n\, H^{2n}\, O^2] + (n - 2)\, CH^2\, O^2$$

Fatty acid. Formic acid.

Formic acid itself would be decomposed into carbon dioxide and hydrogen.

Crude calcium tartrate, mixed with organic matter, and left under water, in summer, ferments and produces an acid which was at first identified with propionic acid, but which, according to Limprecht and Von Uslar, must be isomerous with ordinary propionic acid.

Crude tartar, without the addition of lime, gives nothing but acetic acid. We have—

$$\underset{\text{Tartaric acid.}}{2\ C^4\ H^6\ O^6} = \underset{\text{Propionic acid.}}{C^3\ H^6\ O^2} + \underset{\text{Carb. diox.}}{5\ CO^2} + \underset{\text{Hydrogen.}}{H^6}$$

$$C^4\ H^6\ O^6 = \underset{\text{Acetic acid.}}{C^2\ H^4\ O^2} + 2\ CO^2 + H^2$$

In the fermentation of mucic acid, of which we have spoken above, the butyric acid, which only appears slowly, and in small quantity, ought to be considered as a secondary product. The principal phenomenon may be summed up in a chemical point of view by the equation—

$$C^6\ H^{10}\ O^8 = 2\ C^2\ H^4\ O^2 + 2\ CO^2 + H^2$$

All the compounds of the malic group, such as malic, fumaric, aconitic, aspartic acids, and asparagine, undergo, under the form of salts of lime, and in the presence of animal matter in process of decomposition, a transformation into different products, among which we may consider succinic acid as the principal term.

With malic acid, we should have—

$$\underset{\text{Malic acid.}}{2\ C^4\ H^6\ O^5} = \underset{\text{Succinic acid.}}{C^4\ H^6\ O^4} + \underset{\text{Acetic acid.}}{C^2\ H^4\ O^2} + \underset{\text{Carb. diox.}}{2\ CO^2} + \underset{\text{Hydrogen.}}{H^2}$$

The sum of the last three terms of the second member

gives the composition of tartaric acid; we may, therefore, consider succinic fermentation as the result of two reactions. In the first, two molecules of malic acid are converted into one molecule of tartaric acid, and one of succinic acid:

$$\underset{\text{Malic acid.}}{2\ C^4\ H^6\ O^5} = \underset{\text{Succinic acid.}}{C^4\ H^6\ O^4} + \underset{\text{Tartaric acid.}}{C^4\ H^6\ O^6}$$

The second reaction would be the fermentation of tartaric acid formulated above.

The production of valerianic acid in the succinic fermentation of calcium malate is explained by the equation—

$$\underset{\text{Succinic acid.}}{2\ (C^4\ H^6\ O^4)} = \underset{\text{Valerianic acid.}}{C^5\ H^{10}\ O^2} + 3\ CO^2 + H^2$$

As maleic, fumaric, and aconitic acids only differ from malic acid by less water, their transformation into succinic acid is explained in the same manner.

As aspartic acid may be considered as amido-succinic acid, and asparagine as amido-succinamic acid, the production of succinic acid at their expense is not surprising.

We have just passed in review a great number of reactions; they all belong to the class of fermentation, because they are excited by certain bodies which act only by their presence.

They have, besides, one thing in common—the formation of acids of the fatty series.

We must consider it as yet vague and uncertain whether the cause of these fermentations is to be attributed to the presence of a special ferment for each, as we have seen in lactic and alcoholic fermentations.

In fact, this cause has only been thoroughly studied and determined by Pasteur, for the butyric fermentation of sugar and calcium lactate.

In the fermentation of ammonium tartrate, M. Pasteur found a ferment similar to lactic ferment, which, acting on ammonium paratartrate, causes the decomposition of the levo-tartrate ; which affords a new example of elective fermentation, comparable to that which we have observed in levulose and glucose.

In M. Pasteur's opinion, the butyric decomposition of calcium lactate is due to the presence in the liquid of a special ferment—*fermentum butyricum.*

"Butyric ferment is composed of little cylindrical rods, rounded at the extremities, usually straight, either isolated or united in a chain of two, three, or four joints, and even of more. The diameter of these small rods is generally $\frac{2}{1000}$ of a millimètre, and the length of the isolated portions from $\frac{2}{1000}$ to $\frac{20}{1000}$ mm. (·0000687 to ·000687 in.). These organisms move forward by sliding. During this movement their body remains rigid or undulates slightly ; they spin round, they balance themselves on end, and agitate their extremities: they are often bent. These singular organisms are reproduced by fission.

"The butyric ferment is, therefore, an infusorium of the genus vibrio." (Pasteur, Comp. Rend., 52, p. 344, February, 1861).

The same writer has ascertained that this ferment, placed in a solution of sugar containing phosphates and ammoniacal salts, reproduces, and causes butyric fermentation.

The conditions of its development are similar to those of lactic fermentation.

The most favourable temperature is 40° C. (104° F.): the medium should be neutral, or slightly alkaline. An acid medium is opposed to the development of the germs of butyric fermentation. However, when once formed, they can live and excite the decomposition of sugar or lactic acid in an acid medium, provided there be no excess of acidity.

M. Pasteur at first asserted that butyric vibrios not only lived without free oxygen,* but that oxygen kills them. The respiratory theory of fermentation, proposed by this observer, does not agree with this fact. If fermentation is the result of such a need of oxygen, that the ferment takes it up from organic compounds, exciting their decomposition by a rupture of equilibrium, we cannot understand how oxygen can act as a poison to the ferment. I do not know whether M. Pasteur has since maintained the opinion that oxygen kills butyric vibrios.

The conditions of nutrition of butyric ferment are, according to Pasteur, the same as those of ferments in general. However, considering its tardy appearance, as compared with lactic ferment, in mixtures which undergo lacto-butyric decomposition, we may admit that it requires albuminoid substances in a process of more advanced change for its nourishment.

In sugar, we generally see viscous, lactic, and butyric fermentation appear in succession. However, it is not proved that butyric fermentation cannot appear before lactic fermentation, and thus act on the sugar itself.

Putrefaction—Putrid Fermentation.—Albuminoid substances, and bodies allied to them, which enter into the

* The alcoholic and lactic ferments are in the same category.

composition of living organisms, have, for a long time, enjoyed a special reputation for instability, which varies, however, according to the nature of the substance. Before Pasteur's researches, it was generally admitted, that as soon as the influence of life is withdrawn (the vital force which alone was able to maintain them in their integrity), these products begin to be transformed, to change, and to be decomposed into several principles, among which are found compounds with a strong and putrid odour.

The remarkable researches of Appert on the preservation of animal substances, and those of Gay-Lussac on the fermentation of the must of grapes, had given the idea that the momentary intervention of oxygen is necessary to excite the first step in this decomposition. This initial impulse once given, the phenomenon of decomposition goes on spontaneously; and the organic matter in process of transformation is even susceptible of transmitting the molecular movement with which it is imbued to more stable bodies, such as sugar, which of themselves undergo no modification. This is, as we have already seen, the theory of fermentation borrowed by Liebig from Stahl and Willis, with certain modifications of form.

However, Schwann (Ann. de Poggend., 41, p. 184), Ure (Journ. Prakt. Chem., 19, 186), and Helmholtz (J. F. Prakt. Chem., 31, p. 429), had shown that the greater part of bodies subject to decomposition, when heated in a retort with water, so as to drive out all the air by boiling, are no longer decomposed, if instead of allowing ordinary air to enter the retort as it grows cold, we are careful only to admit air previously subjected to a red heat. Under these conditions, putrefaction does not make its appearance, and we no longer observe the development of infusoria and mildews.

The abundant presence of infusoria and mildews in putrefaction had been long known, but it was not thought that these microscopic beings were the true causes which determined the decomposition. They are developed, it was said, owing to germs brought by the air, or already contained in the decomposing bodies, or by spontaneous generation, and because they find a soil favourable to their nutrition. The bond between their appearance and putrefaction was only one of concomitance.

Schwann, and the other authors quoted above, thought, on the contrary, that the germs of infusoria and of mildews set up putrefaction by their development; and, as a proof of this, they brought forward their experiments, in which putrefaction was no longer produced, when the pre-existent germs were destroyed, and their introduction by means of the air was prevented.

As the calcination of the air gave rise to some objections, especially to that of a possible change in the constituent principles of this gaseous mixture, Schroeder and Th. V. Dusch (Ann. der Chem. und Pharm., vol. 89, p. 232), repeated Schwann's experiments, with this difference, that instead of allowing calcined air to re-enter the retort in which the organic substance had been boiled, they simply filtered the air through a sufficiently thick layer of cotton wool; they thus succeeded in mechanically arresting the germs and solid matters held in suspension, but without influencing in any way the properties and the composition of the air. The wort of beer, broth, meats recently boiled in water, are then preserved very well, even during the heat of summer.

Some contradictory facts, however, lent support to the arguments of the adversaries of the theory of putrefaction under the influence of infusoria. Thus, the authors of the before-mentioned experiments had themselves ascertained that milk recently boiled coagulates, grows sour, and putrefies, just as well in air that has been strained as in ordinary air ; meat not steeped in water, but simply heated in a water-bath, is also not preserved in strained air : in these two cases we observe neither infusoria nor mildew, and yet decomposition is produced.

It is then evident, it was said (Gerhardt, Chimie Organ., vol. 4, pp. 545), that it is indeed the air which brings and deposits in matters in a state of putrefaction the germs of organisms, but it is not less certain that these are not the first cause of decomposition, since it can be produced without their intervention. If the calcined or strained air is less active, in many experiments, than ordinary air, it is because not only are the germs of infusoria removed by these operations, but also the remains of decomposed matter which are suspended in it ; that is to say, ferments whose activity would be added to that of oxygen.

The question was in this state when Pasteur resumed the study of putrefaction, by looking upon it in the same light as had guided him in his researches on fermentation. Sustained by the idea that all these phenomena can be explained by the presence, the development, and the multiplication of microscopical plants or animals, he sought to prove that some of these exist also in the putrefaction of animal nitrogenous substances.

These experiments were conducted by two methods which lead to the same end, and confirm each other.

On the one hand, they tend to show that putrefaction is always accompanied by the presence, the development, and the multiplication of infinitely small, organized, living beings; on the other hand, they prove, that whenever we place ourselves under conditions calculated to avoid the presence of the germs of organisms, at the commencement of the experiment, decomposition does not take place, even in *products the most liable to it.*

By their precision and their extent, they are calculated to remove the objections raised by the partial decompositions noticed by his predecessors, Schroeder and V. Dusch, and which we have mentioned before.

We will speak of the second series of experiments when we treat of the origin of ferments, only saying here, that the many trials made by Pasteur lead to a positive solution of the question. By preventing contact of the germs with animal matter, we prevent, at the same time, every trace of fermentation from showing itself.

M. Pasteur distinguishes two orders of phenomena in putrefaction; some are produced under the influence of organic ferments which live without the aid of oxygen, like butyric ferment; in others, on the contrary, the oxygen takes part, as an essential element, promoting combustion; oxidation is also excited by organisms.

We shall treat of slow combustions in the chapter on acetic fermentation; this will give us a simple and clear example of the reactions of this group, and may be considered as the type of slow combustions. We have nothing to add here to what will be said on this subject, and there only remains for us to speak of putrefactions without oxygen, or *putrid fermentation.*

When, in a putrescible liquid, containing albuminoid organic matter, the dissolved oxygen has been absorbed, and has completely disappeared under the influence of the first infusoria developed, such as the *Monas crepusculum*, and the *Bacterium termo*, "the *vibrio* ferments, which do not require this gas to sustain their life, begin to show themselves, and putrefaction is immediately set up. It is accelerated by degrees, following the progressive increase of the vibrios. As to the putridity, it becomes so intense, that the examination of a single drop of the liquid, under the microscope, is a very painful task."

"It follows, from what has been said, that contact of air is by no means necessary for the development of putrefaction. On the contrary, if the oxygen dissolved in a putrescible liquid was not at once removed by the action of special organisms, putrefaction would not take place; the oxygen would destroy the vibrios which would try to develop at first."

When the putrescible liquid is exposed to the air, we notice the two kinds of reactions simultaneously; there forms on the surface a complete film, composed of bacteria, mucors, and mucidines, which excludes the oxygen, and prevents its penetrating into the liquid. The vibrios which multiply there, under shelter of this rampart, transform by fermentation the albuminoid matter into more simple products, while the bacteria and mucors excite the combustion of these products, and bring them back to the state of the least complex chemical combinations. Such is the representation of the whole of the phenomena of putrefaction as drawn by M. Pasteur (Comp. Rend., June, 1863).

We have already made some reservations relative to this singular property of vibrios—that of not being able to endure the presence of oxygen.

The opinion held by Schwann, Ure, Helmholtz, Schroeder, and V. Dusch, and finally by Pasteur, relative to the cause of putrefaction, is corroborated by the very process which is employed to preserve perishable bodies. The conditions of preservation are precisely such as oppose the development of organisms.

Such are the employment of cold, zero C. (32° F.), and below; also of a sufficiently high temperature. Cooked albuminous matter resists putrefaction much longer, because the germs which were there are destroyed; but decomposition will, nevertheless, show itself if we do not carefully guard against effects from without. Appert's process, which consists in cooking meat, or other perishable substances, in iron boxes hermetically sealed, realizes these conditions. The germs are killed, and there is no possibility of fresh ones entering. As, at the same time, the small quantity of air contained in the box loses its oxygen, it has been thought that the preservation depended on this complete elimination of the oxygen at 100° C. (212° F.).

The total absence of water very efficaciously opposes the development of living organisms. Thus we can preserve, as we may say, indefinitely, dried meat and vegetables.

All substances known as antiseptics are also enemies to ferments. Thus common marine salt, alcohol, creosote, phenol, salicylic acid, sulphurous acid, the sulphates, potassium acetate (Sacc.), carbon dioxide, tannin, the acids, many metallic salts, as those of copper, mercury, iron, aluminum; potassium chromates, arsenious acid, prussic acid, lime water, the antiseptic properties of which are well known and have been frequently tried, all these are also poisons for ferments of various kinds in the quantities in which they are active.

The preservative action of oil, grease, ashes, fine sand, bran, sawdust, coatings of paraffin or gelatine, is explained by these porous or impermeable bodies preventing the approach and access of germs brought by the air, like the cotton wool in Schroeder's experiment.

Products of Putrefaction.—The products of putrefaction are very numerous. This may be easily understood, first, because the putrid change of an organ or liquid directly taken from the animal or vegetable economy is the resultant of the decomposition of the various constituents which are found in it. The special study of the products of putrefaction of each particular albuminoid substance has only been attempted in a very few cases.

In the second place, the compounds, definite in appearance, which undergo putrid fermentation, are so complex in their constitution, that we ought to expect to meet with a great number of derivatives formed by putrid decomposition.

The most constant products which make their appearance in putrefactions screened from the air are leucine, and probably some of it homologues, tyrosine, the volatile fatty acids of the series $C^n H^{2n} O^2$ (formic, acetic, propionic, butyric, valerianic, caproic, &c.), ammonia, and some compound ammonias (ethylamine, propylamine, amylamine, trimethylamine), carbon dioxide, sulphuretted hydrogen, hydrogen, and nitrogen.

If we refer to what will be presently said with reference to the decomposition of albuminoid substances under the influence of barium hydrate, we shall be able easily to account for the appearance of these various products. On one hand, the albuminoids contain the elements of urea, and ought to be considered as compound ureids. This fact alone explains the appearance of carbon dioxide, and of a part of the ammonia. (*See* ammoniacal fermentation.)

The albuminoids are decomposed by hydratation under the influence of baryta, furnishing leucine and some of its homologues, tyrosine and a sulphide. These first products may, probably, undergo the ulterior action of ferments, and yield ammonia and volatile fatty acids. We know, in fact, that in presence of putrefied fibrin, leucine is resolved into ammonia and valerianic acid.

$$C^6 H^{13} N O^2 + 2 H^2 O = C^5 H^{10} O^2 + N H^3 + CO^2 + H^4$$

Everything leads us to believe that putrefaction is a complex phenomenon—that it is only a successive series of fermentations exerted on more and more simple products.

Thus, for example, when we leave fibrin to spontaneous decomposition, without access of air, it is resolved into two principles, as under the influence of sea-salt. One of these principles is albumin, which, on account of its greater resistance to the action of ferments, will be found for a long time in the putrid liquid. The second product of this decomposition, undergoing somewhat quickly a more thorough change, yields acetic, butyric, valerianic, and capric acids, as well as ammonia (Brendecke), which are evidently derived from the amido-acids homologous with leucine.

The chemical reactions which accompany the putrefaction of the albuminoids are, then, for the most part, phenomena of hydratation, which may be reproduced identically by chemical forces alone, independently of vital action. Thus we shall see that phenomena of this kind may be excited by the action of soluble ferments, whether diastasic or indirect; and we are induced to suppose that a part, at least, of the transformations undergone by proteids, and their more immediate derivatives, are the consequences of phenomena of this order (indirect fermentation).

Nothing resembles putrid fermentation, with reference to the derived products, more nearly than the change which takes place in the constituent parts of yeast, when left to itself without nourishment, deprived of sugar and oxygen.

We see, in fact, the appearance of leucine, tyrosine, sarcine, &c. This is the first step; the action stops there, and goes no farther; the yeast, or the special soluble ferment which it secretes, is unfit to attack these bodies again; but if we wait for the development of vibrios, we shall find the production of ammonia, carbon dioxide, and volatile fatty acids, at the same time that the leucine partly disappears.

M. Ulysse Gayon has published quite recently, as a thesis for the "Doctorat ès Sciences" (Paris, 1875,

Faculté des Sciences de Paris, No. 362), the result of many experiments on the spontaneous decomposition of eggs. The question was important, and very interesting to the adversaries of the theory of spontaneous generation. Besides, the facts observed by M. Donné and M. Béchamp on this subject seemed contrary to the ideas of M. Pasteur on the general cause of putrefaction. M. Gayon, a pupil and demonstrator of M. Pasteur's, endeavoured to bring the spontaneous decomposition of eggs and their putrefaction under the general law enunciated by his teacher.

M. Donné (Expériences sur l'Altération spontanée des Œufs, Comp. Rend. de l'Ac. 57, p. 450, 1863) had said: "If we take eggs in their natural state, not shaken, and leave them to themselves, they remain for weeks and months, even during the great heat of summer, without undergoing any putrid decomposition. The egg has no unpleasant smell, and nothing, either possessing animal or vegetable life, is produced, either on the surface of the membrane or in the inside; there are no traces of infusoria or microscopical vegetation.

"If, on the contrary, we destroy the physical structure of the interior of the egg by shaking; if, that is to say, we break up the texture, and the cells of the albuminous substance, and thus mix together the yolk and the white, then, even without access of the external air, and even guarding against this intervention by extra precautions, such as a coating of collodion spread over the surface of the egg, we find all the phenomena of decomposition make their appearance, after a longer or shorter time, according to the temperature, but always in less than a month; all the phenomena of decomposition, with the exception, however, of the production of living organisms, either vegetable or animal; for, whatever may be the degree of rottenness to which we allow the egg to proceed, we can never discover the slightest trace of animalculæ, or of microscopic vegetable life; the matter of the egg grows troubled, and of a livid colour; it exhales a fetid odour directly we break the shell, but nothing, absolutely nothing, stirs in its substance; nothing lives, and the most careful and frequently repeated examination by means of the microscope does not enable us to discover the least trace of an organized or living being."

We may add, that M. Béchamp also found no organisms in rotten eggs.

M. U. Gayon's experiments, into the details of which we cannot enter, led him to the following conclusions:—

"Putrefaction in eggs, whether in the presence or the absence of air, is correlative to the development and multiplication of microscopical organisms of the family of *vibriones*.

"In other terms, contrary to the result found by M. Donné and M. Béchamp, eggs make no exception to the great law of correlation which M. Pasteur has demonstrated for all the phenomena of fermentation, properly so called."

We are thus, on the subject of eggs, confronted by two distinct affirmations, as much opposed as black and white. M. Donné found them; M. Gayon did not. We have no balance wherewith to estimate and compare the skill of the two observers. It appears to us certain that M. Gayon saw what he described; but we cannot affirm that M. Donné was absolutely mistaken, and that the eggs, in the conditions under which he placed them, contained vibrios which he did not find.

In the absence of any other criterion, we bring forward a very important fact, mentioned by M. Gayon himself.

This skilful microscopist observed that some of the eggs experimented upon at the temperature of about 25° C. (77° F.), whether shaken or no, underwent a special modification, distinct from ordinary putridity and from acid fermentation.*

The decomposed mass is of a dirty yellow colour, it has an odour of dried animal matter, and is very fluid; we see in it, also, a great number of needle-like as well as botryoidal crystals, formed of tyrosine. It contains much greater quantities of tyrosine and leucine than are found in ordinary putrefaction. M. Gayon was not able to discover, under these circumstances, any trace of microscopic organisms, either in the inside, on the surface, or in the substance of the membranes. However, tyrosine and leucine are evident and unquestionable symptoms of the decomposition of albuminoid matter.

Between the production of these substances and the

* In certain cases, M. Gayon saw the contents of eggs, especially of those which had been shaken, transformed into a homogeneous mass, of the consistence of butter, and of a bright yellow colour, with a sour smell, and a strongly acid reaction. M. Béchamp observed a similar change in an ostrich egg, and was able to distinguish the presence of alcohol, acetic acid, and sulphuretted hydrogen. M. Gayon attributes this change, to which he gives the name of acid fermentation, to the presence of special organisms. These are small immovable rods, with pale outlines and homogeneous tints, either isolated or articulated two and two together, from $\frac{5}{1000}$ to $\frac{10}{1000}$ of a millimètre in length, and $\frac{5}{10000}$ to $\frac{7}{10000}$ of a millimètre in thickness.

phenomena called putrefaction there is, chemically speaking, no very clear distinction to be drawn. They are reactions of the same order, decompositions, more or less extensive, of the proteid molecule; the traces of sulphuretted hydrogen, and other fetid products which communicate such a repulsive odour to putrefaction, cannot serve to establish an absolute and philosophical line of demarcation between the decomposition without organisms observed by M. Gayon, and what is wrongly termed putrefaction properly so called.

The result of this seems to be, that albuminoid matters are able to undergo certain decompositions, certain changes, without the intervention of living organisms.

By means of a very simple and ingenious apparatus, M. Gayon succeeded in extracting the gas contained in large ostrich eggs in a state of putrefaction.

One of these eggs, in a state of thorough putrefaction, yielded 150 cubic centimètres of gas, containing per cent.:—

Sulphuretted hydrogen	Traces.
Carbon dioxide	30·5
Hydrogen	40·2
Nitrogen	29·3
	100·0

The presence of nitrogen might be due to the accumulation of a certain quantity of air in the air-bubble before putrefaction.

Among the solid and liquid products of the putrefaction of eggs, the presence of small quantities of leucine and tyrosine, alcoholic products, and volatile acids (butyric acid) were recognized. The sugar had disappeared.

CHAPTER XII.

FERMENTATION BY OXIDATION.

ACETIC fermentation and the reactions which we shall class with it under the generic name of fermentation by oxidation, have a special character, which we have not met with in any of the phenomena which we have hitherto studied.

Not only are the fermentable matter and the ferment concerned in the reaction, the one furnishing the constituent parts of the new bodies which are formed, and the other acting as a cause, but we find that a third factor, the oxygen of the air, becomes necessary.

In other words, under the name of fermentation by oxidation, we shall speak of combustions set up by living organisms, which serve as media between the oxygen of the air and the combustible body or fermentable matter. As to the results of this combustion, they may vary, according to the nature of the body which is burnt, and may even be resolved into the simple products of the most complete combustion (water and carbon dioxide).

We will begin with the *acetic fermentation of alcohol.* It has been long known that the alcohol contained in fermented liquids, such as wine, beer, &c., will disappear under certain circumstances, and give place to vinegar or acetic acid, and that the air, or rather its oxygen, plays a part in this reaction. The progress of chemistry, and the exact determination of the respective composition of alcohol and acetic acid, give a simple and clear account of the reaction; it may be formulated thus:—

$$\underset{\text{Alcohol.}}{C^2 H^6 O} + O^2 = H^2 O + \underset{\text{Acetic acid.}}{C^2 H^4 O^2}$$

The oxidation may take place by two reactions, with the production of an intermediate product, aldehyde:—

$$\underset{\text{Alcohol.}}{C^2 H^6 O} + O = H^2 O + \underset{\text{Aldehyde.}}{C^2 H^4 O}$$

$$\underset{\text{Aldehyde.}}{C^2 H^4 O} + O = \underset{\text{Acetic acid.}}{C^2 H^4 O^2}$$

In a purely chemical point of view, we have here a very simple phenomenon, on which we need not dwell.

The determining causes of this oxidation will have greater claim on our attention. Döbereiner having shown by an experiment, which has become classic, that the vapour of alcohol mixed with the oxygen of the air becomes acid, being transformed into acetic acid, thought that he had ascertained by this means the true theory of the phenomenon, the essential conditions of which became the simultaneous action of alcohol and oxygen, in presence of a porous body, such as finely divided platinum, charcoal, wood shavings, &c., capable of favouring by a *catalytic* action, that of contact, the oxidation of the alcohol.

It was on this idea that the process of rapid acetification, called the German process, was founded; this was first tried by Schützenbach in 1823. Long before this, Boerhaw had already introduced into practice an analogous commercial process. He employed vats three mètres high and one and a half mètres in diameter, with a double bottom pierced with holes, placed at 30 centimètres (about a foot) from the bottom. Bunches of grapes, from which the juice had been expressed, were placed on this false bottom, so as to fill the vat. One of the vats was entirely filled with wine, the other was only half filled ; after twenty-four hours, liquor was drawn from the full vat and poured into the other so as to fill it ; this operation was alternately repeated till the acidification was complete. It is seen that, in the vat which is half full, the alcoholic liquor which soaks the grape stalks and skins is exposed to the action of air on a large surface, and the oxidation is singularly favoured by this fact.

In Schützenbach's process, use is made of a large oak vat, from two to three mètres in height by one in diameter, furnished with a false bottom pierced with holes, and placed about 30 centimètres from the bottom. Some centimètres higher, the circumference of the vat is regularly pierced with a series of holes passing entirely round it. These orifices are inclined from without inwards, so as to prevent the liquor from escaping.

At the upper part, at the distance of 30 centimètres from the lid, another false bottom is placed, pierced with many small holes and several large ones ; the latter are usually closed by plugs, and are intended for the purpose of renewing the air, when it is deoxygenated. The whole is closed by a cover furnished with an opening carrying a funnel which can be closed, and serves for the introduction of the liquor to be oxidated. The interval between the two partitions is filled with beech-shavings. A thermometer placed in the interior of the vat gives an idea of the intensity of the reactions.

All being thus arranged, hot vinegar is first poured into the vat; this filters through the shavings, impregnates them, and serves afterwards to facilitate, or rather to set up, the oxidation of the alcohol. In order to be transformed into vinegar, suitable mixtures of alcohol and vinegar are employed, similar to those which are used in the Orleans process. The temperature of the surrounding air is maintained at 21° C. (about 70° F.). The alcoholic liquor employed is heated to between 26° and 27° C. (79° and 81° F.); the temperature rises spontaneously to 38° to 42° C. (101° or 108° F.) in the vats.

The alcoholic liquor is never completely oxidized by its first passage through the vat; the operation is once or twice repeated, either in the same vat, or in others by the side of it. (For further details *see* special works on chemical technology.) In good manufactories, a result is obtained which does not differ by more than six per cent. from that indicated by theory; and even this loss may be lessened by certain suitable accessory arrangements, to which we need not now allude.

The French method of acidification of wine, called the Orleans method, is very different. We will say a few words about it before we enter on the results of Pasteur's researches, and the commercial consequences which have been derived from them.

The oxidation takes place in barrels placed side by side, on wooden frames, supported by stone pillars. At the upper front part of each vat two orifices of unequal size are made; the larger serves for the purpose of introducing and drawing off liquor, the other for the admission of air. These vats, which contain from two to four hundred litres (44 to 88 gallons) called *mothers*, are at first one-third filled with strong boiling vinegar; from 11 to 12 litres (20 to 21 pints) of wine are then added, and the vats are then left undisturbed; at the end of a week's acetification, another quantity of wine is added, and so on from time to time till the vessel is half full of vinegar. A third part of the contents of the "mother" is then drawn out by means of a siphon, and wine is again added from time to time in portions of 11 to 12 litres (20 or 21 pints). A regular and continued process is thus carried on. It is often found that, without any apparent cause, a certain "mother" refuses to acidify the wine, or only gives rise to a very slow and tedious oxidation; in this case, the employment of a much stronger wine, or a much higher temperature, sometimes restores the action to its normal activity. These anomalies have only been explained by Pasteur's experiments. The operation in one barrel is considered to be terminated when a stick plunged into the liquid is covered with a thick white froth (flour of vinegar); as long as the froth is red, the addition of wine is continued. The most favourable temperature is between 24° and 27° C. (76° to 82° F.).

In this special manufacture of vinegar, the influence of porous bodies in determining the oxidation cannot be brought to bear, as in the experiment with platinum or in the German method of manufacture. The vinegar-makers, from their special experience, attribute the acidification to the action of a deposit which forms in the barrels, and to which they give the name of „ mother of vinegar."

According to the opinions of Liebig, which have so long prevailed in scientific matters, either dead or living organic matter, when in contact with alcohol in wine, possess the property, after having absorbed oxygen, of oxidizing, at the ordinary temperature, organic and inorganic substances. This property would therefore exist in solid organic substances which are in a state of decomposition or putrefaction. To support this view, Liebig (Ann. de Chim. (4), vol. 23, p. 178,) recalls de Saussure's experiment, who found that soil placed in a mixture of hydrogen and oxygen gave rise to the formation of water and the disappearance of hydrogen, and of an equivalent quantity of oxygen. We may state in a few words, and without entering into too many details on a question which seems determined, that Liebig explains the production of acetic acid by a certain movement communicated by principles in process of decomposition —a movement which in this case excites oxidation.

In fact, his ideas are wanting in clearness. Sometimes he seeks for the key to the phenomenon in a catalytic action of porous bodies; sometimes, with reference to his general theory of fermentation, he attributes it to the influence of a ferment (organic matter in process of decomposition).

Such was the state of the question when Pasteur undertook his researches on the causes of the spontaneous acidification of wine and alcoholic liquors.

According to him, the oxidation of alcohol is the consequence of the action of a cryptogam of the genus *Mycoderma*. We will give a summary of the experiments on which his opinion was founded.

If on the surface of any organic liquid, necessarily

containing phosphates and nitrogenous organic matter, we allow any species of mycoderma to develop itself, until the whole surface of the liquid is covered with it; if then we carefully remove the nutritive liquid by means of a siphon, without suffering any portion of the membranes to break up; then, if we substitute for this liquid an equal volume of water, containing 10 per cent. of alcohol, we immediately see the plant which is placed under these abnormal circumstances of nutrition, set up a reaction between the oxygen of the air and the alcohol of the liquid. The acetification begins immediately, and goes on with great activity. After a certain time, the action, impeded by the great acidity of the liquid, proceeds more slowly; but we can restore to it all its activity by substituting alcoholized water for the acid liquid. A time, however, comes when the plant, becoming partly decomposed itself, communicates to the liquid, in consequence of the organic and mineral elements of its dead tissues, properties which serve as nutriment for the various species of mycoderma. The action then takes on a different phase; the acetic acid and alcohol disappear with great rapidity, and the liquid becomes completely neutralized; because as soon as the plant finds in the subjacent medium the nutritive principles suited to its development, it sets up much more intense oxidizing action, and burns, not only the alcohol, but also the acetic acid, converting it into water and carbon dioxide.

This complete combustion is noticed whenever we cause the mycodermata to be developed on alcoholic liquids containing food fit for the nourishment of the plant, such as wine, beer, or fermented organic liquids;

unless, however, we place the mycoderma, whether intentionally or unintentionally, under the conditions of incomplete or tardy development, that is to say, in a sickly state.

To sum up, we may say that the mycodermata, developing on the surface of an alcoholic liquid which contains suitable nutritious principles, burn the alcohol, and bring it to the same state as oxygen at red heat, that is, complete combustion. If, on the contrary, we diminish the vital activity of the mycoderma, whether by depriving it of its nourishment or by any other means, the oxidizing action which it may be able to set up will not go so far, and the alcohol may change into acetic acid. M. Pasteur's experiments will also show that the influence attributed to ordinary porous organized bodies, in the German manufacture of vinegar, is the result of imperfect observation.

The beech-shavings act, not on account of their porosity, but because their surface is covered with thin pellicles of mycoderma ; the many points of contact with the air favour the action, but are not the determining causes of it. To prove this, M. Pasteur caused some alcohol diluted with water to trickle down a cord. The drops which fell from the end of the cord did not contain the smallest quantity of acetic acid. The experiment lasted for more than a month, the liquid trickling extremely slowly, only one drop in two or three minutes. If we repeat this experiment, having previously steeped the cord in a liquid on the surface of which there is a pellicle of mycoderma, a portion of which clings to the cord as it is withdrawn, the alcohol, which passes slowly down this cord, will be charged with acetic acid, and this acetification may be continued for several weeks.

M. Mayer made some similar experiments, which entirely confirmed Pasteur's results. Thus filter paper, previously boiled with hydrochloric acid at 50 per cent., then with caustic soda at 50 per cent., and finally washed with distilled water, was laid on the surface of an alcoholic liquid, without giving rise, even after the expiration of a month, to the least trace of acetic acid. The result was also as negative when a funnel was filled with fragments of glass and of the same paper, and alcohol at 9 per cent. was allowed to run slowly through this filtering mass.

M. Pasteur did not go more deeply into this question, and did not ascertain by what means the mycodermata could thus give rise to more or less energetic oxidations.

In one of his latest writings on this question (Réponse aux Critiques de M. Liebig, Ann. Chim. Phys. (4), vol. 25, p. 148, 1872), he thus expresses himself:—

"This little microscopical plant (*Mycoderma aceti*) possesses the power of condensing the oxygen of the air in the same manner as spongy platinum or blood-globules,* and of conveying this oxygen to matter beneath."

It seems, from these few words, that M. Pasteur compares the action of his organic ferment, the *Mycoderma aceti*, to that of spongy platinum: his theory of acetic fermentation really differs very slightly from that

* The oxygen of the globules of blood is fixed in the hæmaglobin, and its oxidizing power is scarcely more energetic than that of ordinary dissolved oxygen.

of Liebig. The latter admits that inanimate porous substances participate in the properties of spongy platinum; Pasteur, on the contrary, attributes this quality only to living organisms.

Mayer's experiments tend to prove that oxidation by mycodermata is a special biological action, which cannot be attributed solely to the physical condition of the plant which acts as ferment. In fact, we have only to heat an alcoholic liquid, covered with its pellicle of acetic mycoderma in full process of acidification, and we shall arrest all oxidation; yet it is difficult to believe that under these conditions the physical state has been sensibly modified.

M. Mayer also noticed striking differences between the mode of action of spongy platinum and that of the *Mycorderma aceti.* For the first, an elevation of temperature above 35° C. (95° F.) favours the oxidation, by augmenting the tension of the vapour of alcohol, while the maximum activity of the *Mycoderma aceti* is placed between 20° and 30° C. (68° and 86° F.); its oxidizing power is destroyed under 10° C. (50° F.) and above 35° C. (95° F.). We can oxidize very concentrated alcohol with platinum; the concentration is even a favourable factor; while for physiological acetification, the alcohol employed ought scarcely to contain more than 10 per cent. of ethyl hydrate.

The immediate and practical consequence of the results obtained by Pasteur is, that in order to act efficaciously, the mycoderma ought to be in contact, at the same time, with the air and the alcoholic medium. The new commercial method for the acetification of fermented liquids, now employed to some extent at Orleans, is founded on this principle. The following is the process :—

The *Mycorderma aceti* is first sowed on the surface of an aqueous liquid containing 2 per cent. of alcohol, 1 per cent. of vinegar, and traces of alkaline and alkalin-earthy phosphates. When the surface is covered with the membrane, the alcohol begins to acidify. This action being fully set up, some alcohol, wine, or beer mixed with alcohol, is added every day to the liquid, in small quantities; this is continued till the oxidation becomes slower; the acetification is then allowed to terminate, and the vinegar is drawn off. The membrane is collected, washed, and employed for a new operation. It is better always to give the plant sufficient alcohol, so that its activity should not be exerted on the acetic acid. Nor ought it to remain too long out of the liquid, or it would lose its active force; finally, it is better to moderate its development, to prevent burning oxidation.

A vat, one square mètre in section, and containing from 50 to 100 litres (from 11 to 22 gallons), can furnish per day from 5 to 6 litres (9 to 10½ pints) of vinegar. The successive phases of the operation are ascertained by means of a thermometer divided into tenths of a degree, the bulb of which is plunged into the liquid.

When we act upon diluted alcohol, it is better to add to the liquid $\frac{1}{10,000}$ of a mixture of magnesium phosphate, and potassium and ammonium phosphates.

Special arrangements allow the introduction of the liquid, without there being any necessity for disturbing the superficial film of mycoderma. The vessels are cylindrical, or of a prismatic form, of one square mètre in section, and $\frac{1}{5}$ mètre in depth, closed by a cover furnished with orifices for the admission of the air.

FIG. 22.—Mycoderma aceti.

It only remains for us to say a few words about the botanical character of the *Mycorderma aceti*.

The continuous membranes, either wrinkled or smooth, which are found on the surface of liquids while in acetic fermentation, are generally formed (Fig. 22) of very minute elongated cells, whose greater diameter varies from 1, 5, to 3 thousandths of a millimètre (·000059 to ·000118 in.); these cells are united in chains, or in the form of curved rods. Multiplication seems to be effected by the transverse division of the fully developed cells. This division is preceded by a median constriction, which has been considered by some authors as a morphological characteristic of the cell.

It appears, from this description, that the mycoderma of vinegar belongs to the family of bacteria.

The general conditions of the nutrition of acetic bacteria have been discovered by Pasteur, and closely resemble, up to a certain point, those of beer-yeast.

Thus mineral salts, alkaline and alkalin-earthy phosphates, proteid nitrogenous substances, or ammoniacal salts, are elements necessary for the devolopment of these organisms. Diluted alcohol (10 per cent. at most) seems, in this case, to take the place of hydrocarbon matter; it may be supplemented by the acetic acid; for, according to M. Pasteur, the progressive weakening which vinegar undergoes, when the acetification is left too long to itself, is due to a subsequent burning of the acetic acid.

M. Blondeau (Comp. Rend., vol. 57, p. 953) has even observed that sugar can acidify without passing through the condition of alcohol, under the influence of the mother of vinegar.

We must, however, notice that the activity of the ferment is increased by the presence in the liquid of a certain quantity of acetic acid.

Antiseptic agents, in general, which by their presence delay and arrest the development of the yeast of beer, and consequently alcoholic fermentation, act in the same manner with respect to the *Mycoderma aceti.* Sulphurous acid is especially active in this manner; and it is partly in order to avoid the acetification of wine, that care is taken to put in the tuns intended to receive it, sulphurous acid, or to burn sulphur matches in them.

The *Mycoderma vini,* of which we have spoken above (p. 59), resembles in many respects the acetic ferment. Like it, it is developed on the surface of fermented alcoholic liquids, in the form of smooth or wrinkled films or membranes; but these latter, however, are thicker and more compact. It acts also as the means for conveying the oxygen of the air to the alcohol of the medium, and to the other combustible principles; but, under its influence, combustion is complete, and accompanied by the production of carbon dioxide and water; it is to this action that we must attribute the rapid weakening of wines covered with mycoderma. We have elsewhere seen that, when immersed in the midst of a saccharine liquid, it can act like the yeast of beer, and produce alcoholic fermentation.

Its nutritive principles are the same as those of the mother of vinegar (alcohol, salts, nitrogenous compounds); besides, it appears also capable of utilizing for nutrition certain secondary products of alcoholic fermentation, such as succinic acid and glycerin. The forms of the cells of this mycoderma—forms that we know to be variable, seem to depend in a great degree on the conditions of nutrition.

Its activity of development appears included between 16° and 30° C. (61° and 86° F.).

Slow Combustion.—The *Mycodermata* of which we have just spoken are not the only organized ferments capable of exciting the slow combustion of carburetted materials.

It has long been known that organic matter, of vegetable or animal origin, left in contact with air, undergoes progressive and complex transformations, known under the name of putrefaction, of slow combustion, and of eremocausis, whose effect is to transform them into principles more and more simple, by means of decomposition and oxidation; so that, in the end, the carbon is restored to the air in the form of carbon dioxide, the hydrogen under the form of water, the nitrogen either as free nitrogen or ammonia. M. Pasteur has distinguished, among the complicated facts of putrid fermentation, two orders of distinct phenomena, although each is connected with the reactions set up by living organisms. The first includes the putrefaction, which takes place without the assistance of oxygen in the air, which is caused by the presence of vibrios. We have spoken of this in the chapter on butyric fermentation, with which these phenomena are connected.

The second, slow combustion, is due to bacteria, mucors, and mucidines, that is to say, to vegetable ferments, which, like the *Mycoderma vini*, and others, possesses the remarkable property of exciting the oxidation of a great number of organic principles, such as sugars, alcohols, organic acids, albuminoid nitrogenous matter, &c., at the expense of the oxygen of the air.

After having proved, by careful experiments, to which we will return when we treat on the origin of ferments, that spontaneous slow combustion of animal or vegetable substances depends necessarily on the development of organisms in the interior, or on the surface, of the substances which are in process of decomposition, and that without organisms there is neither combustion nor absorption of oxygen, M. Pasteur traces the following picture of putrid decomposition in contact with air (Comp. Rend., June, 1863):—

"Even the most easily decomposed animal matter, as, for instance, blood or urine, may be preserved for an indefinite length of time in air which has been calcined or deprived of its germs; under these conditions, the absorption of oxygen is but trifling, and putrefaction does not take place; and, at the same time, no infusoria are produced. If, on the contrary, this same substance remains exposed to the ordinary air, it is oxidized, putrefies, and infusoria are developed.

"It is commonly known that putrefaction takes a certain time to declare itself, a period varying according to the circumstances of temperature—of the neutral, acid, or alkaline character of the liquid. Under the most favourable circumstances, at least twenty-four hours are required before the phenomenon begins to manifest itself by external signs. During the first period, an internal movement takes place in the liquid, the effect of which is to withdraw entirely the oxygen of the air which is in solution, and to substitute for it carbon dioxide gas. The total disappearance of the oxygen, when the medium is neutral or slightly alkaline, is generally due to the development of the smallest kinds of infusoria, especially the *Monas crepusculum*, and the *Bacterium termo*. A very slight troubling then takes place, because these little beings pass about in all directions. If the vessel containing the putrescible liquid has a large opening to the air, the bacteria perish only in the liquid mass, after the removal of the oxygen, while they continue, on the contrary, to propagate, *ad infinitum*, on the surface, because it is in contact with the air. There they cause a thin film to form, which goes on thickening by degrees, until it falls to the bottom of the vessel; then another forms, and so on continually. This film, to which different mucors and mucidines are attached, prevents the solution of oxygen gas in the liquid, and consequently allows the development of vibrios. With respect to these latter organisms, the vessel is as if it were closed against the introduction of air.

The putrescible liquid thus gives rise to two very distinct kinds of chemical action, which have reference to the two sorts of organisms which are nourished in it. On one hand, the vibrios, living by the co-operation of the oxygen of the air, set up in the interior of the liquid acts of fermentation—that is to say, they transform the nitrogenous matter into more simple, but still complex, products.

The bacteria (or the mucors), on the other hand, consume these same products, and bring them to the state of the most simple ordinary combinations, water, ammonia, and carbon dioxide.

The compounds which longest resist slow combustion are the fixed fatty acids, forming the adipocire of the old chemists, cellulose or its derivatives formed by dishydratation, such as ulmic acids, vegetable mould, peat, &c.

The oleic acid, on the contrary, disappears altogether. But little is known of the details of these various phenomena of slow combustion.

Can the same organism set up the combustion of different organic principles, differing from one another in their constitution? Is the action of the oxygen progressive, or is it complete from the very commencement? These and many other secondary questions, more or less interesting, present themselves; but their solution requires long and minute researches, similar to those which relate to alcoholic fermentation. However, the cause of the phenomena is known, and the way for new investigations is open.

CHAPTER XIII.

APPLICATIONS OF THE RESEARCHES AND IDEAS OF M. PASTEUR.

WE have already seen, when treating of acetic fermentation, what consequences M. Pasteur has drawn from his observations, in order to regulate and facilitate the transformation of fermented liquors into vinegar.

The manufacture of beer may also, according to this investigator, derive advantage from the careful study of fermentation and of ferments.

If we add to the wort of beer ordinary yeast, principally composed of cells of *Saccharomyces cerevisiæ*, with very few foreign organisms, under the most favourable conditions for the development of yeast, it will reproduce almost alone, setting up direct alcoholic fermentation. As the quantity of original yeast is found, after the fermentation, to be six or seven times greater than at first, supposing that no foreign germs have been introduced from without, we can understand that the new yeast will be purer than the first. By continuing with this, new fermentations of the wort of beer, we shall procure, by a kind of selection similar to that described by M. Raulin, with reference to *Aspergillus niger*, a very pure yeast, free from the mixture of other organisms.

When this result has once been obtained, it will be sufficient to maintain the integrity and purity of ferment, by excluding from contact with air the fermenting vats of beer, and by carrying on the process in closed vessels, instead of open vats. The principle of the new process patented by M. Pasteur, into the details of which we cannot enter, depends then on the employment of pure yeast, and on fermentation with the exclusion of air, in order to avoid the introduction of foreign organisms, which, by their ulterior development, might produce changes of another order, lactic fermentation, &c.

For this purpose, the wort, after it is prepared, is drawn off while boiling into vessels made of wood or metal, and cooled in a current of carbon dioxide, or of air purified from ferments, and then yeast is added.

The beer, after the first fermentation, is drawn off into casks, in which it ripens and grows clear. The wort may be carried to a great distance, and the beer has, according to M. Pasteur, superior qualities both as to taste and preservation.

Preservation of Wine.—M. Pasteur has made a long and very attentive study of the changes which may take place in wines in the various phases of their preservation. His observations have been preserved in a very excellent work published on this subject. We can only give here the more general results obtained by this author. He attributes the changes which take place in wines to the development of special living ferments.

A special organism corresponds to each order of change, *i.e.*, to each disease of the wine. The germs of these ferments are found in the must of fermented grapes,

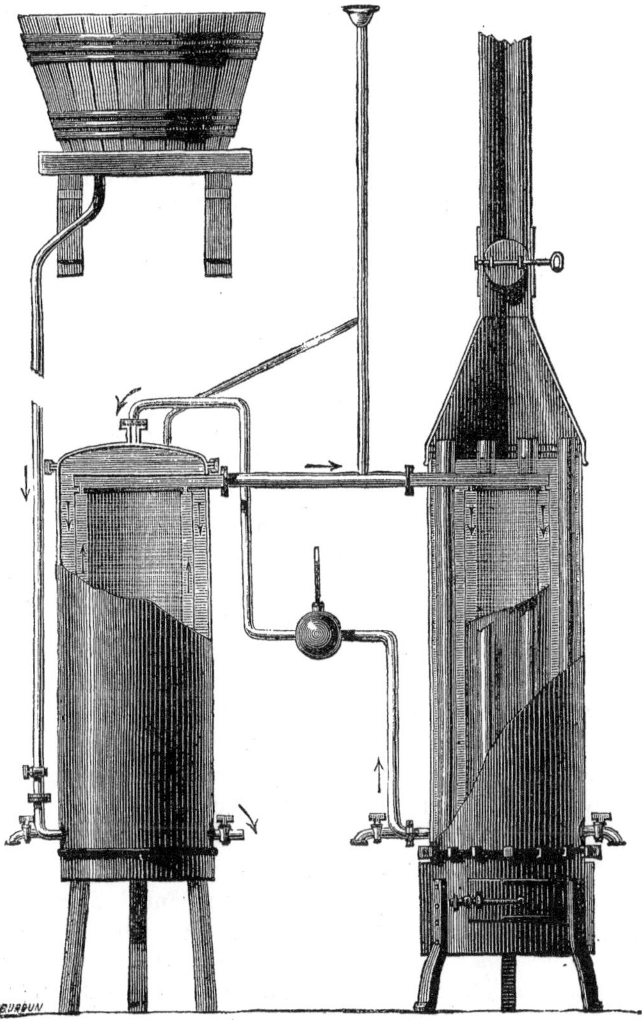

FIG. 23.—Giret and Vinas' apparatus for warming wines.

and may develop themselves whenever the conditions become favourable.

In order to prevent their development, and to allow the wine to be kept for an indefinite term, without any change, we have only to raise the temperature for a moment to 60° C. (140° F.). The germs are killed at this comparatively low temperature, because of the presence of alcohol (from 8 to 10 per cent.).

The warming of wines for the purposes of preserving them had already been proposed by M. Vergnette de la Motte, but the rational explanation of this process is due to M. Pasteur. He also determined with greater precision the conditions of the operation.

This warming may be effected while the wine is in bottle or in casks.

Fig. 23 represents the apparatus of Giret and Vinas for the continuous warming of wine. It is composed of a stove surmounted by upright tubes communicating with the chimney. A water-bath completely surrounds these tubes. The cylinder of the water-bath is fixed to the stove by means of two rims, between which is a slip of cloth saturated with paste. The two rims are compressed by iron clamps, so that this cylinder can be easily removed. The wine circulates in a cavity formed by two concentric cylinders, secured above and below by two annular collars. The refrigerator is formed of a cylinder containing an interior cavity similar in every respect to the preceding one. The cover of the refrigerator is movable, and is fastened by an arrangement like that which serves to connect the cylinder with the stove.

The surfaces in contact with the wine are tinned. The

water-bath contains water, the refrigerator, as well as the enclosure around it, only holds wine.

The arrows show the direction of the circulation ; a thermometer placed within the bulb of the tube which connects the water-bath with the refrigerator gives the maximum temperature. (For further details on this subject *see* " La Chimie Technolique de Wagner," French translation.)*

Application of M. Pasteur's ideas to Pathology.—Some years since I wrote in the following terms (La Chimie appliquée à la physiologie animale à la pathologie et au diagnostic médical, par P. Schützenberger, 1864) :—

" All diseases, contagious either by inoculation or by more or less direct contact, whether epidemic or endemic, are evidently produced by the introduction of foreign poisonous substances into the living organism, producing true poisoning. When in an affection of this kind, as for example, in cholera, yellow fever, or malignant ulcers, the general symptoms and the mode of evolution have a well-marked constancy of character, notwithstanding differences of race, species, and individuals, one is compelled to admit the specific nature of the poison which gives rise to certain pathological manifestations.

" These conclusions, drawn from many facts, observed in every part of the world, are so simple and natural that no one denies them ; but when it is required to state the precise nature of the morbific influence ; as we enter the domain of hypothesis, the most various and contradictory opinions have been given, and may be supported with a greater or less appearance of probability.

"Infectious diseases were, for a long time, attempted to

* There is an English translation, by W. Crookes, London.

be explained by intra-organic fermentations, set up by foreign bodies which everything induced us to consider as of an organized nature; but at a time when the ideas of fermentation, properly so called, were vague and ill-defined, it was difficult to maintain with certainty such a doctrine.

"It has long appeared to us that Pasteur's researches on this subject have not only succeeded in determining with greater precision than before facts up to that time partly known, but that they are destined in the future to throw a bright light on etiology, and on the pathological history of contagious diseases, whether epidemic or endemic. Yet we should not have ventured to discuss here a personal conviction, which, though shared by many physicians and observers, only rested on analogy, were it not that a recent discovery, due to the skilful investigation of M. Davaine, had corroborated this opinion by a positive fact scientifically determined, a fact which will certainly not remain unsupported by others.

"We set out with the hypothesis that the greater part of infectious diseases have for their immediate cause the penetration into the organism, and the development of the living germs of ferments, or living ferments already formed, of either animal or vegetable nature, and we will make use of the knowledge already acquired in order to support this opinion with a certain amount of probability. We must premise, however, that it will be necessary, before we decide this question definitively, to wait for a direct and experimental proof, such as M. Davaine has given with reference to the "blood of the spleen" or malignant boils. We think that researches on fermentation and putrefaction have been carried far enough to allow important trials to be made in this direction with some hope of success."

During the ten years since this page was written, skilful experimentalists, guided by the same ideas, among whom I may mention Pasteur himself (researches during the cholera epidemic), have studied this subject with great care, and yet I must admit that there has been no result from these inquiries ; the question of the etiology of infectious diseases has made no important advance; the observation made by M. Davaine remains without any additional support.

Are we then to conclude that these attractive predictions are erroneous, and must be rejected, or that these observations are impossible, even with the increased microscopical power which we possess? It is difficult to decide one way or the other, and every question requires for its decision positive facts; negative results can only serve as a means of checking our observed facts.

BOOK II.

ALBUMINOID SUBSTANCES—SOLUBLE OR INDIRECT FERMENTS—ORIGIN OF FERMENTS.

CHAPTER I.

ALBUMINOID SUBSTANCES, OR PROTEIDS.

ALBUMINOID substances, or proteids, play so important a part in biological phenomena in general, and in the nutrition of ferments in particular, that it would be impossible for us not to devote a few pages to the study of these bodies.

If these substances represented chemical species well defined and correctly classed among organic compounds—in other words, if their constitution were thoroughly known, we should merely refer the reader to works on pure chemistry, as we have already done in the case of fermentable sugars. But unfortunately, in spite of many important investigations, it must be said that the history of proteids is still one of the most obscure subjects of organic chemistry, one of the most urgent desiderata of biological science.

So long as the question of the constitution of the immediate principles of animal tissues has not been determined, it will be in vain for physiological chemistry to investigate by the most careful direct analysis the different elements of an organ, or of a liquid, whether in a normal or pathological condition; we shall always be stopped by the unknown in the interpretation of the results which have been obtained. The many and various reactions which take place in the organism may be regarded as true fermentations, in which the fermentable bodies are partly represented by proteids. It is not necessary to go into this at great length to make it understood how much these reactions, as observed in the different active phases of an organ, will gain in scientific value, when we are able to formulate them by means of an equation, as we give the formula of alcoholic fermentation.

We shall study albuminoid substances, by taking this general view of them, and by bringing out more prominently, among the results obtained, those which throw some light on the constitution and mode of decomposition of these bodies.

The existence of a certain number of albuminoid substances, considered as distinct species, has been admitted, being founded on more or less important differences in their physical or chemical properties, or in their elementary composition. This division and classification may be made by two methods. If we notice only very decided differences of character and composition, we may form out of albuminoid substances different groups or families united by common bonds, but sufficiently distinct for confusion or error to be rendered impossible.

Among these groups many closely allied species may be placed, which owe their existence and individuality only to certain subtile divergencies of properties which present to the mind nothing very distinct, such as the manner of forming precipitates with different reagents.

Without entering into these details, we will take as types of each group the most important principle from a biological point of view. It cannot be said that substances to which a special name has thus been given, as the albumin of the egg, casein, and fibrin, are immediate well-defined principles; for here especially we have no criterion by means of which we are able to establish a chemical species. We may, indeed, suppose, with a certain amount of probability, that they are only mixtures, in variable proportions, of bodies very nearly allied and almost identical, the separation of which from each other would be very difficult, if not impossible. The natural fatty bodies give us instances of this complicated combination of products very similar in their composition and character, and which direct analysis has scarcely any power to separate.

This hypothesis has also derived some support from exact observations. Thus, albumin, which was long considered as an immediate principle, is, in fact, only compounded of many albumins, having *very nearly* the same composition, and which can only be distinguished from each other by their rotatory power, and by the temperature at which they coagulate.

This question can only be settled by a careful study of the products of the analysis and decomposition of albuminoid substances; as the nature of fatty bodies has only been understood by examining the products of their saponification, which are more easily separated than are the original bodies.

These views, which are verified more and more, were developed by M. Bouchardat (Thès pour le concours d'agrégation, 1872), and by M. Berthelot, after a communication made to the "Société Chimique" by the author of this book. They are very reasonable, and capable of wide application.

The nitrogenous principles which enter essentially into the constitution of the organs and liquids of the animal and vegetable economy are naturally divided into various families.

The first includes albuminoid substances, properly so called; that is to say, bodies the most nearly allied, by their chemical composition and by the whole of their properties, to the albumin of egg.

The second is formed of more remote products, usually less rich in carbon and containing more nitrogen. They enter into the composition of less vital tissues, that is to say, those in which the phenomena of nutrition and the changes are more restricted; of those which do no work, that is, do not develop much force, such as bony, cartilaginous, elastic, fibrous, cellular, horny, or epidermic tissues. In this family we find horny tissue, or keratrin, ossein, epidermose, elasticin, mucin or the gelatines, and the fibroin of silk.

The soluble ferments, the peptones, are products of the decomposition of albuminoid substances, and should be classed separately.

We find in the organism, besides these, certain mixed principles placed between hydrocarbons and proteids, true complex nitrogenous glucosides, decomposing into glucose and other nitrogenous principles, such as chondrin and chitin.*

Albuminoid Substances, properly so called.—Into this family we generally admit the following compounds:—

1. Albumin of the egg, the albumin of serum and (serine), vegetable albumin; these substances are soluble in water.

2. Casein, vegetable casein, paralbumin, syntonin, myosin, legumin, amandin, the proteids or albuminates of the German chemists (this name cannot be adopted in French, since it gives an idea that the body in question is a salt), paraglobin, metaglobin or fibrinoplastic or fibrinogenous substances.

3. Blood-fibrin, coagulated albumin, amyloid matter, and vegetable fibrin.

With regard to their centesimal composition, albuminoid substances are very similar to each other; yet the analyses do not agree sufficiently to allow us to conclude that their composition is identical, or that they are isomeric. We subjoin in a tabular form the principal analytical results:—

* According to my latest researches, albuminoid substances, even those of the first family, contain, in small proportions, cellulose amides as an integral part of their molecule. The distinction established between the albuminoids and chitin is therefore not absolute. The latter contains, it is true, a greater proportion of cellulose amide.

	Carbon.	Hydrogen.	Nitrogen.	Oxygen and Sulphur.	Sulphur.	Authors.
Egg albumin, not coagulated	53·3	7·1	15·8	23·6	1·8	Dumas and Cahours.
Egg albumin, not coagulated	54·3	7·1	15·7	22·9	—	Scheerer.
Egg albumin, purified ...	52·9	7·2	15·6	—	—	Wurtz.
Egg albumin, coagulated	52·9	7·2	15·8	—	—	Wurtz.
Blood fibrin	52·8	7·0	16·8	23·4	—	Dumas and Cahours.
" "	53·7	7·1	15·8	23·4	—	Scheerer.
Casein	53·5	7·1	15·8	23·6	—	Dumas and Cahours.
"	54·0	7·2	15·7	23·1	—	Scheerer.
Gluten, or vegetable fibrin	53·1	6·8	15·0	—	—	Boussingault.
Legumin	53·7	7·2	15·7	23·4	—	Scheerer.
"	50·5	6·9	18·2	—	—	Dumas and Cahours.
Amyloid matter... ...	53·6	7·0	15·0	—	1·3	—
Syntonin	54·1	7·2	16·1	—	1·1	—

On the one hand, it is difficult to attach too great importance to small differences, observed in analytical results obtained with bodies so difficult to purify, amorphous, and often containing mineral substances; on the other hand, nothing authorizes us to admit the complete identity of their composition.

In fact, these bodies have certainly a very high molecular weight. Thus for albumin, by two methods, (analyses of the potassic and platino-hydrocyanic combinations) nearly the same number, 1612, has been arrived at for the molecular weight; so that if we wish to translate the results of the elementary analysis into a chemical formula, as is usually done, we are led to very high numbers in the expression as in that proposed by Lieberkuhn, C^{72} H^{112} N^{18} S O^{22}.

But it is evident to any one who is in the habit of calculating analyses, that a difference of one atom of carbon, hydrogen, or oxygen, in a formula with such high numbers, gives variations which are within the limits of the errors of analysis. Elementary analysis, as we are at present capable of performing it, cannot solve the question of isomerism or non-isomerism.

Are we farther advanced in that which relates to the constitution of these bodies? At present, we can foresee that, though the question is still in suspense, it will not be long before it is determined.

In order to ascertain this constitution, it is necessary to know exactly the whole of the constituent parts resulting from the destruction of proteids, under certain determinate conditions. But in the various reactions which cause their splitting up and transformation into more simple principles, we have been able to recognize the formation of certain well-defined compounds; but these bodies are more frequently accompanied by a relatively considerable number of uncrystallizable bodies, which have not yet been studied, and which render any attempt to determine an equation of the reaction illusory.

On this account, we cannot form any definite idea of the manner in which the 72 atoms of carbon, the 112 atoms of hydrogen, &c., of the albumin are united.

I will give a short summary of the principal results obtained with respect to these reactions.

As long since as 1820, Braconnot observed the production of the sugar of gelatin or glycocoll, $C^2 H^5 NO^2$ (amido-acetic acid) by boiling gelatin in sulphuric acid moderately diluted; by substituting muscle for gelatin he obtained, under the same conditions, the leucine, $C^6 H^{13} NO^2$ (amido-caproic acid).

Liebig afterwards showed that another crystallizable

product is formed at the same time, tyrosine, $C^9 H^{11} NO^3$ (oxyphenyl-amido-proprionic acid).

Erlenmeyer and Schaeffer, by extending these researches (the action of boiling diluted sulphuric acid), to most of the albuminoid substances, observed the constant formation of leucine and tyrosine.

They obtained for 100 parts of dried matter—

	Leucine.	Tyrosine.
Fibrin	14	·8
Albumin	10	1·0
Syntonin	18	1·0

Casein gives leucine and tyrosine, and a sirupy residuum.

Ritthausen obtained by the same kind of reactions, from products formed by the action of boiling dilute sulphuric acid on vegetable nitrogenous substances, such as gluten, two well defined crystallizable acids.

Coaglutin yielded an acid of the formula $C^5 H^9 NO^4$ (glucamic acid), homologous to aspartic acid.

Legumin gives, under the same circumstances, legumic acid, $C^8 H^{14} N^2 O^6$.*

According to Hlasiwetz and Haberman, the greater part of animal and vegetable proteids are capable of producing these nitrogenous acids (aspartic, glutamic), when boiled with dilute sulphuric and hydrochloric acids.

I have lately been led to study, with great care and in all its details, a reaction which allows the albuminoids to be almost entirely resolved into crystallizable principles. My first experiments were made in order to ascer-

* Ritthausen says that he has since found that legumic acid is only a mixture of glutamic and aspartic acids. I do not think that this opinion is correct.

tain whether a part of the nitrogen of proteinic compounds was not found in the state of urea, and whether this class of bodies does not represent complex ureids.

After having vainly sought for urea among the products of the physiological decomposition of proteids, while ferment was kept without nourishment, I boiled albumin, casein, &c., with barium hydrate, in an apparatus so arranged as to allow the ebullition to continue during several successive days, without any diminution of the water. I could easily accomplish this, by heating it in a retort, attached to a condenser so placed as to allow the water to return; the latter was connected with two Wolfe's bottles, containing a known volume of normal sulphuric acid, intended to retain the ammonia disengaged.

Under these conditions at 100° C. (212° F.), it was found that during the first few hours there was an abundant disengagement of ammonia, which diminished by degrees, and at last became imperceptible.

Nasse had previously observed this fact, and had for this reason, divided the nitrogen of the albuminoid substances into two portions; one, the least important in quantity would be found, as he stated, feebly fixed or combined (*Losegebundener Stickstoff*).

The barium experiment gives us the key to this difference. In fact, at the same time that the ammonia is set free, we see a granular precipitate formed in the liquid which was originally clear; this precipitate increases progressively to a certain limit, and then ceases to be formed.

It is almost entirely composed of barium carbonate, mixed with a small quantity of barium sulphite and oxalate, with silica derived from the glass attacked by the alkaline liquid, and with alkalino-earthy phosphates contained in the albuminoid matter used in the experiment.

By measuring the *ammonia* and *barium carbonate* formed, after boiling for 120 hours, there were found in 100 parts by weight of dry albumin—

	Gram.
Ammonia	1·7
Barium carbonate . . .	11·1

By calculating the weight of carbon dioxide which corresponds to the barium carbonate, we find that the quantities of carbon dioxide and ammonia set free, are almost exactly in the proportion which urea would yield, by resolution into ammonium carbonate:—

$$CH^4 N^2 O + H^2 O = CO^2 + 2 NH^3$$

$$\text{ratio of } \frac{CO^2}{2 NH^3} = \frac{44}{34} = 1{\cdot}29.$$

We have, then, reason to admit the presence of the same grouping of elements as in urea in albuminous substances.

Boiling at 100° C. (212° F.), even prolonged for eight days, does not entirely destroy this grouping ; but albumin is resolved, under the influence of barium hydrate, into several more simple combinations, possessing various degrees of stability.

When, however, the temperature is raised to 140° or 150° C. (284° or 302° F.), the decomposition is complete in a few hours (12 to 24), and the quantities of ammonia and carbon dioxide remain the same, even if we prolong the operation, and raise the temperature of the mixture to 200° C. (about 391° F.). These experiments must necessarily be made in a close vessel.

Under these conditions, we find in 100 parts of dry albumin—

Ammonia disengaged	4·2	4·5
Barium carbonate	25	29

These numbers give also the ratio of 1·29 between CO^2 and NH^3.

Albumin must therefore contain, in 18 atoms of nitrogen, about 4 belonging to the urea group.

M. Béchamp has noticed, among the products of the oxidation of albumin by potassium permanganate, the presence of a small quantity of urea. This fact, which was denied by the German chemists, has been confirmed by Ritter's experiments. However this may be, the urea found existed in but small quantity in the products of the reaction; and it is possible that its production is the result of decomposition, making that which is already formed sensible rather than the result of an oxidation.

The barytic liquid, separated from the ammonia and from the barium carbonate, and freed from the baryta in excess by a current of carbon dioxide, retains but little colour; by concentration and crystallization, followed by treatment of the mother-liquor with alcohol, we succeed, at last, in bringing it entirely into a crystalline form, that is to say, to the state of a definite body. We may, therefore, by making a direct analysis of the crystals obtained, arrive at a complete notion of the composition of albuminoid substances. The reaction of the barium hydrate, at from 150° to 200° C. (302° to 394° F.) gives us an opportunity of trying every analytical method on the products resulting from the splitting up of albuminoids. These are not merely two or three defined and crystallized bodies, which are separated from a mass of sirupy matter of considerable quantity, which still remains undetermined; on the contrary, we get all the compounds that are formed, and we thus find ourselves in a position to construct an equation of the composition of albumin and allied substances.

Although my researches on this subject are not entirely completed, I may already state, from the results obtained,—

1. That barium hydrate decomposes albuminoid substances by simple hydratation, the experiment being made in the absence of oxygen.

2. That the principal products of this reaction are—*the elements of urea* (ammonia and carbon dioxide, in the proportion of 1·29); traces of sulphurous acid, of sulphuretted hydrogen, and of oxalic and acetic acids; *tyrosine* $C^9 H^{11} NO^3$ (oxyphenyl, amido-propionic acid), in very small quantity, two or three per cent. of the albumin at the most.

We also find,—

3. The *amido acids* of the series $C^n H^{2n+1} NO^2$, corresponding to the fatty acids $C^n H^{2n} O^2$, from amido-œnanthylic acid $C^7 H^{15} NO^2$ to amido-propionic acid; leucine $C^6 H^{13} NO^2$; butalanine $C^5 H^{11} NO^2$; and amido-butyric acid $C^4 H^9 NO^2$ abound in this mixture.

4. *One or two acids* nearly allied to *aspartic* and *glutamic* acids $C^5 H^9 NO^4$ and $C^4 H^7 NO^4$; *one or two acids* analogous, and very nearly similar, to the legumic acid found by Ritthausen $C^8 H^{14} N^2 O^6$.

A small quantity of a substance analogous to dextrin, which, by being boiled in acids, is converted into a body which energetically reduces Fehling's liquid; nitric acid changes it into oxalic acid.

With the exception of these bodies, nothing of importance is found either as a chemical compound or as an amorphous mass.

Most of the definite compounds found by this reaction are similar to those which have already been mentioned among the products of the splitting up of albuminoid substances, under the influence of acids; but I repeat that the importance of the preceding reaction consists especially in the demonstration that these compounds alone constitute the albuminoid molecule; and in the proof that the elements of urea or carbamide form an integral part of this molecule.—(*See* the Memoir of the author on Albuminoid Substances, Bulletin de la Soc. Chim. de Paris, February 15th, March 5th, and March 15th, 1875.)

The albumin is not split up all at once in this manner into comparatively simple products; by stopping the reaction at different periods of its development, or by using occasionally less elevated temperatures, we meet with intermediate compounds, uncrystallizable, or crystallizable with difficulty, the thorough study of which will be very interesting, in relation to the phenomena of nutrition and of biological reactions.

The products which we obtain under the influence of the alkalies melted in their water of crystallization (ammonia, hydrogen, ammoniacal compounds, methylamin, saniline, picoline, petinine, leucine, tyrosine, glycocoll, carbon-dioxide, formic, valerianic, butyric, and oxalic acids) are derived from a more energetic action, exerted on the first series of compounds formed by the action of baryta. It is the same for the compounds which originate under the influence of oxidizing agents, such as the mixture of dilute sulphuric acid and potassium bichromate (formic, acetic, butyric, valerianic, caproic, and propionic acids, with their corresponding aldehydes, benzoic acid and benzoil hydride, hydrocyanic acid and butyl cyanide).

It does not enter into our programme to write the complete history of albuminoid substances. (The reader may consult with advantage on this subject the Dictionary of Chemistry, by M. Wurtz; l'Histoire Générale des Matières Albuminoides, par G. Bouchardat; Thèse pour l'Agrégation, Paris, 1872.) We shall merely give a summary of the theories proposed, in order to explain the composition of these complex bodies.

All albuminoid substances, heated for some time with alkalies, dissolve and give up sulphur to the alkali, under the form of sulphide and hyposulphite; the neutralization of the liquid by an acid causes a voluminous floccose, white precipitate, soluble in dilute lye-water; this precipitate has the same composition whatever may be the albuminoid matter employed. Mulder, relying on these facts, considered this precipitate as the base of albuminoid substances, and gave it the name of *protein*. The German chemist thought at first that his protein no longer contained sulphur; later experiments have shown that in reality it still contains some, but that this sulphur cannot be removed by alkalies under the form of sulphides. According to Mulder's theory, all albuminoid substances are combinations of protein with variable quantities of sulphur, phosphorus, and mineral matter.

This opinion, admitted very generally at first, was by degrees abandoned, on account of the great number of contradictory observations.

Liebig considered albuminoid substances as having the same elementary composition, *i.e.*, as isomeric compounds. This view, which admits of discussion, throws no more light than that of Mulder on what is now called the chemical constitution of these compounds, considering that protein is almost as complex in its character as the original products. Sterry-Hunt gave an explanation which at first sight was very seductive; he considered albuminoid substances as amides or nitrites of cellulose, dextrin, gum, or sugar.

The objection to this simple idea is, that it does not agree with known facts; tyrosine, leucine, and aspartic acid are not considered as derived from sugar, from cellulose, or from their nitrites. These latter bodies (nitrites of hydrocarbon substances) have been also little studied, and are little understood.

Berthelot, in his treatise on elementary chemistry (1872), considered, from the whole of the facts then known, albuminoid substances to be complex amides, formed by the union of amido acids of the series $C^n H^{2n+1} NO^2$(such as glycocoll and leucine), of tyrosine, with certain oxygenated principles, some of which belong to the acetic series, and others to the benzoic series.

He thought that the nature of the amides, and of the oxygenized bodies which generated them, as well as their relative proportions, was the cause of the differences which exist between the various albuminoid bodies; and that chitin and chondrin contained, in addition, the elements of glucose.

The results which I have obtained by means of the decomposition by baryta, show that the albuminoids are formed by the association, in different proportions, of the *urea* and *amido-acid combinations*, some belonging to the series of leucine, $C^n H^{2n+1} NO^2$, the others, which are more highly oxygenized, belonging to the series $C^n H^{2n-1} NO^4$ (aspartic and glutamic acids); the more complex acids, such as the legumic, may be considered as the products of an incomplete decomposition. Tyrosine ($C^9 H^{11} NO^3$) represents the aromatic series; it is from this that benzoic and paroxy-benzoic acids and bromanil are derived, being obtained under different conditions.

Besides this, it is easy to see that, with the amido-acids $C^n H^{2n+1} NO^2$, it is impossible, even when without the urea, to arrive at the composition of the albuminoids; whatever combination of these bodies we may choose, when we deduct the elements of water in sufficient proportion to obtain the value of the oxygen of the albumin, there remains a very notable excess of hydrogen. The intervention of the amido-acids of the aspartic series $C^n H^{2n-1} NO^4$, in the grouping of proteids, is therefore indispensable to explain the constitution of these bodies.

CHAPTER II.

SOLUBLE FERMENTS AND INDIRECT FERMENTATION.

IN the chemical phenomena which we have hitherto studied, we have found the cause of the reaction connected in such an intimate manner with the presence of a cellular organism, that all efforts made with the view of separating this cause from this form of life, even for an instant, have failed, or at least have produced no precise result definitely adopted by the scientific world. The farther progress of science will, we hope, allow us to penetrate more deeply into the very essence of the phenomena.

If the remarkable and important labours of Pasteur have taught us that the transformations of sugar into alcohol, lactic, and butyric acid, gum and mannite, those of the albuminoid substances into various putrid principles, as well as the conversion of alcohol into acetic acid, depend on the presence of inferior organisms, and that the germs of these organisms come from without, it is no less true that we have as yet no certain idea of the mode of action of organic ferments. There have been, it is true, hypotheses enough to bring about the solution of the problem. Liebig, obliged to allow more than he had hitherto done to the presence of living organisms, said in his last memoir (Ann. Chim.

Phys. (4), vol. 23, p. 6), that, in a chemical point of view, the only one which he will not give up, a "vital act," is a phenomenon of motion, and that in this sense M. Pasteur's opinion is not contradictory to his own, and is not a refutation of it.*

We have already discussed elsewhere the theory of M. Pasteur, propounded as early as 1861, and lately resumed with greater confidence.

Fermentation is a consequence of the life of ferments without oxygen. These simple organisms need oxygen so much, that when they are in a medium which is deprived of it, they take it even from sugar and other analogous bodies. Fermentation is therefore a consequence of the disturbance of equilibrium resulting from this respiration.

This manner of considering the question is completely at variance with the fact that the production of alcohol at the expense of sugar, under the influence of a ferment, is not interfered with by the presence of oxygen; according to Mayer, it is neither increased nor lessened. Pasteur considers that it is rather increased by oxygen.

It is certain that a ferment is able to live and develop in a saccharine, nitrogenous and mineralized medium, without the intervention of oxygen. These same evolu-

* M. Liebig aims here at the following sentence of M. Pasteur:—"The chemical act of fermentation is essentially a phenomenon correlative to a vital act, commencing and ceasing with it. There is never any alcoholic fermentation without there being, simultaneously with it, organization, development, multiplication of globules, or the continuous consecutive life of globules already formed." According to the German chemist, fermentation is a movement communicated by instable bodies in process of chemical transformation; it signifies but little whether or no these transformations take place in a living organism.

tions are more active under the influence of air. The only inference that we can draw from these well-established facts is, that up to a certain point, the sugar, which the ferment always resolves into alcohol and carbon dioxide, as long as it can meet with it, and unless it is exhausted, is able to supply to the ferment the living forces required for its development. But we know not whether alcoholic fermentation is a consequence of this borrowing of living force, or whether it is a previous action.

The chemical cause of alcoholic fermentation set up by yeast has been sought for from another point of view, which must now be the object of our studies.

We know that cane-sugar, before it undergoes alcoholic fermentation, is hydrated, and thus split up, as often occurs under the influence of acids, into two opposite glucoses; the ordinary glucose, or grape-sugar, which causes the plane of polarization to turn to the right and lævulose, or uncrystallizable sugar. This change was at first attributed to the acidity of the ferment. M. Berthelot, and after him M. Béchamp, showed that the active agent is a soluble, neutral, nitrogenous principle, excreted by the ferment, which is found in greater or less abundance in the filtered water in which yeast has been washed (the zymase of Béchamp, the alterative ferment (*ferment inversive*) of Berthelot).

This soluble principle, to which we can attribute no organization, but which is directly derived from a living organism, possesses the remarkable power of altering cane-sugar in a few moments.

When this fact was established concerning the alterative, soluble, inorganic ferment, analogous compounds had been long known, also soluble, nitrogenous, and inorganic, which were especially characterized by specific acts which they could exercise chemically on various principles.

It was thus that M. Payen and M. Persoz obtained from barley that had germinated, and had been afterwards ground, by treating it with water, a soluble substance capable of turning starch into sugar. The presence of emulsin had also been ascertained in almonds, and this transforms amygdalin into essence of bitter almonds.

That which especially distinguishes these chemical reactions, set up by these various soluble, inorganic principles, is the greatness of the effect compared with the trifling quantity of the active agent. This same character is found also in direct fermentation, due to the immediate intervention of living organisms.

We will give the name of indirect fermentation to the reactions of which we have just spoken, and whose cause is derived from an organism, but can act without it.

It was natural to seek for the cause of direct fermentation in the intervention of active products, whether soluble or not, elaborated by organic ferments. Thus two orders of phenomena, which are certainly not without relation to each other, were connected together.

This theory, which would cause the direct or true fermentation of M. Pasteur to be confounded, as to its essence, with indirect fermentation, or fermentation caused by inorganic soluble ferments, has not found sufficient support from characteristic and well-observed facts, to be established as a truth that has been demonstrated; but it has not yet been absolutely contradicted. It is, however, easily to be seen that such an explanation becomes useless. If the decomposing force is derived from the cell, what would be the utility of this intermediate soluble or insoluble solid substance which transmits the force?

General Characters of Soluble Ferments.—Soluble ferments are all derived directly from living organisms in the midst of which they originate. Up to the present time, the specific characters, of which we shall presently speak, have not been communicated to any artificial organic substance. We are, therefore, compelled to believe that this specific character is a consequence of the origin of soluble ferments. Their composition resembles that of albuminoid substances; in fact, they contain carbon, nitrogen, hydrogen, and oxygen. But the analogy will go no farther. When we have eliminated, by proper processes which we are about to describe, the albuminoid substances which always accompany soluble ferments in their first solutions, we find that the product, though it preserves all its chemical activity, no longer manifests the general reactions of albuminoid substances. It no longer yields a precipitate with tannin and corrosive sublimate; iodine and nitric acid no longer colour it.

Elementary analyses have also revealed sensible differences; but as there was no proof that the products analyzed were pure, we can draw no conclusion from this concerning the chemical nature of soluble ferments. It is probable that they are derived from the physiological splitting up of proteids.

The *soluble indirect* ferments, or *zymases*, present themselves, when in a dry state, as amorphous, colourless, pulverulent matter; they are usually precipitated from their aqueous solutions by alcohol, corrosive sublimate, neutral or basic lead-acetate. These precipitates, when decomposed by sulphuretted hydrogen, restore to the water the soluble matter unchanged, and preserving its specific properties; but it is still, in this case, accompanied by a greater or less quantity of albuminoid matter.

If the sublimate precipitates them from liquids extracted from the organism, it is rather by mechanical action than by the fact of a chemical combination; for we have just seen that soluble ferments, when deprived of albuminous principles, are not precipitated by the sublimate.

By taking advantage of the facility with which the zymases are mechanically carried down by solid deposits in process of formation in the midst of the liquid which contains them, certain chemists have discovered means of purifying them. Thus Conheim separates pure ptyalin (salivary diastase) by acidulating the saliva strongly with tribasic phosphoric acid; the phosphoric acid is subsequently neutralized by lime-water, until an alkaline reaction takes place. The precipitate of tricalcic phosphate carries down with it, as it is formed, the ptyalin and albuminoid matter; the filtered liquid has no action on starch. The deposit, washed with water, yields the ptyalin to the solvent, and retains the proteinic matter; it is then only necessary to precipitate it by alcohol, and we obtain a white, light, flaky deposit, which, dried in vacuo, presents itself in the form of an almost colourless powder.

Pure pepsin is extracted in a similar manner from natural or artificial gastric juice (obtained by exposing the mucous membrane of the stomach, separated from the muscular membrane, and cut up, to a temperature of 35° C. (95° F.), with water containing 5 per cent. of phosphoric acid. This juice, which contains phosphoric acid, is precipitated by lime-water. The tricalcic phosphate is transformed, by the addition of a proper quantity of phosphoric acid, into insoluble and crystallized bicalcic phosphate, which it is only necessary to wash in order to remove the pepsin adhering to it. We may also dissolve the precipitate of calcium phosphate in dilute hydrochloric acid; then pour into the liquid a solution of cholesterin in a mixture of four parts of alcohol and one of ether; shake up with the liquid the cholesterin which is separated, and collect it on a filter; wash it with water acidulated with acetic acid, then with pure water. The damp cholesterin, to which the pepsin adheres, is treated with pure ether, which dissolves it, while there remains in the lower part of the vessel a solution of pure pepsin in water (Brücke). This pepsin yields no precipitate except to platinum bichloride, and neutral and basic acetates of lead. Nitric acid, tannin, and corrosive sublimate are without effect. It only yields a very slight colour with nitric acid and ammonia.

Danilewsky makes use of collodion to precipitate soluble ferments. The precipitate, having been well washed, is treated, after being dried, with a mixture of ether, alcohol, and water, which dissolves the nitrated cellulose, leaving in solution the active principle, free from albuminoid matter.

Von Wittich (Arch. f. d. ges. Physio., vol. 3, p. 339),

proposes the following method, applicable to the extraction of soluble ferments in general. The vegetable or animal organ which contains them is rapidly cut in pieces, cleared from blood, if necessary, by washing in water, and then left under alcohol for twenty-four hours; afterwards dried in the air, pulverized and sifted.

The powder is diffused in glycerin; and this glycerin solution is precipitated by alcohol. By repeating this operation several times (solution in glycerin and precipitation by alcohol) an active powder, free from albuminoid matter, is obtained.

We have but little information respecting indirect ferments, regarded in a chemical point of view. We cannot say whether or no they have the same composition, and only differ by their specific activity. That which gives considerable importance to these products is the transforming power which they exercise over a great number of organic compounds, and the important part which they play in many physiological reactions. The activity of an indirect ferment depends on the temperature, like that of organic ferments. In general terms, we may say that it increases with the temperature up to a certain limit, beyond which it undergoes a rapid depression, till it ceases altogether. This limit varies with the nature of the ferment; it is always under 100° C. (212° F.), and is found to be higher than that of organic ferments.

The action of chemical agents on soluble ferments is also not quite to be compared with that exerted on organic ferments.

Thus M. P. Bert has observed that compressed oxygen destroys the latter ferments after a longer or shorter interval, while soluble ferments are not modified in their activity. This interesting experiment establishes a very distinct line of demarcation between the two kinds of ferments, and may serve to classify them, whenever microscopical indications leave any doubt as to their character.

M. Bouchardat (Ann. Chim. Phys. (3), vol. 14, p. 61), who has carefully studied the influence of chemical compounds on diastase, has observed that certain substances, which are antagonistic to alcoholic fermentation, have no influence on the effects of diastase; such as prussic acid, the mercurial salts, alcohol, ether, chloroform, and certain essences (cloves, turpentine, lemon, mustard, &c.) Citric and tartaric acid, which only slightly interfere with alcoholic fermentation, completely destroy the activity of diastase.

M. Dumas (Comp. Rend., vol. 75, p. 295), made special experiments on the action of borax on this class of ferments. He found that a solution of borax coagulates beer-yeast; the supernatant liquid has lost the property of altering cane-sugar; it also neutralizes the action of the water of yeast on saccharose. If we place sweetened water and the water of yeast in one tube, and sweetened water with yeast-water and a solution of borax in another, the first will soon show signs of alteration, while the second will manifest none. Analogous effects are observed with synaptase or emulsin, diastase, and myrosin. All these ferments cease to act on amygdalin, starches, and myronic acid, from the moment that they are placed in contact with a solution of borax. This salt appears, then, to have a specific action in destroying the activity of all soluble ferments.

We have seen, on the contrary, that yeast, placed in contact for three days with a saturated solution of borax, is able still to set up alcoholic fermentation. Borax may then, like compressed oxygen, serve as a differentiating character of soluble and organic ferments.

Bodies, or Chemical Compounds, on which Soluble Ferments are able to Act, and the Kind of Reactions which they Excite.—Organic ferments exert their activity on a great number of organic compounds belonging to different groups.

The glucoses, acids rich in oxygen, such as malic, tartaric, citric or lactic acids, albuminoid substances, urea, and alcohol, can be affected by organic living ferments. The reactions which these ferments cause them to undergo are often very complex, and cannot be formulated by simple equations, provided that we wish to represent correctly all the terms; in fact, in almost every case, we have not been able to reproduce artificially the complex conditions of organic matter under the influence of organic ferments.

Indirect or soluble ferments are able to act also on various classes of organic compounds, but the mode of action is generally the same. There is a more or less simple splitting up accompanied by a hydratation. The nature of this splitting up is always conformed to the peculiar constitution of the compound, and may be explained, in most cases, by chemical processes in which the direct or indirect intervention of a living organism cannot be brought in.

Thus starch is resolved by hydratation into glucose and dextrin, and this in its turn is converted into maltose, as well under the influence of diastase as by being boiled with a dilute acid (sulphuric acid). The alterative ferment hydrates a molecule of saccharose, and converts it into two molecules of glucose: dilute acids behave in the same manner.

Certain special soluble ferments, such as synaptase or emulsin contained in sweet and bitter almonds, and the myrosin of white or black mustard, act on natural glucosides; that is to say, on those special compounds that are met with in plants in such great abundance, and which we may consider as ethers composed of glucoses, or of polyatomic alcohols. The result is a splitting up, by hydratation, into glucose and another principle. Chemical means, such as boiling with an acid or an alkali, lead to the same result.

The fatty bodies (ethers compounded with glycerin), as is well known by Chevreul's beautiful experiments, fix the elements of water when they are heated with boiling alkalies, or with acids, and yield glycerin and a fatty acid, the sum of the weights of which is equal to the weights of fatty matter used, plus the water fixed by the reaction.

But we know by the works of M. C. Bernard, that the pancreatic gland secretes a soluble nitrogenous substance, capable, like boiling alkalies, of saponifying greasy substances.

It is very probable that the digestion of albuminous matters, and their conversion into peptones, under the influence of the gastric or pancreatic juice, or rather under that of soluble ferments contained in the secretions, are only the results of a hydratation and a splitting up, the conditions of which we are able to realize without the agency of life.

We may almost foresee that, in general, every phenomenon of the splitting up of an organic compound into its proximate constituent parts, requiring only a simple fixation of water, will find in the organism itself its soluble ferment, that is to say, the agent capable of producing this decomposition. It is thus that ethers composed of mon-atomic or polyatomic alcohols, including the glucosides and fatty bodies, certain complex acids, such as the hippuric, glycocholic, taurocholic, &c., are resolved into two or more simpler molecules.

Can the same soluble ferment act upon different chemical compounds, decomposing them in the manner indicated by the peculiar constitution of each? There is not the shadow of a doubt as to the positive answer to this question. Thus we see emulsin or synaptase, contained in almonds, cause the decomposition of a great number of crystallizable and definite principles of the vegetable organism, such as amygdalin, salicin, arbutin, helicin, phlorizin, esculin, daphnin.

It is true that these various bodies belong to the same family, and have the same functions; they are all glucosides, and we must not be more surprised to see the same force transform them than to observe the saponification of neutral fatty bodies under the same influences.

In certain cases, we see the same organic liquid, the same secretion, such as the pancreatic juice, exert its active power upon very different principles, which do not present, as in the previous example, an analogous constitution. The pancreatic juice transforms, modifies, and digests albuminoid substances, turns starch into sugar, and saponifies fatty substances. We may well ask if this multiple specific power belongs to one and the same principle, or whether it indicates the presence of several distinct soluble ferments in the pancreatic juice.

The experiments of Cohnheim and Danilewsky tend to prove that the various physiological properties of the pancreatic juice are due to special principles. These chemists have, indeed, been able to separate by precipitation, by means of calcium phosphate or collodion, a nitrogenous substance, which is not albuminoid, and which shows all the characters of salivary diastase, and which rapidly turns starch into sugar, without digesting the proteids.

Thus, when we pour the ethereal alcoholic solution of collodion into the extract of the gland, obtained by pounding it with magnesium carbonate and water, and afterwards filtering it, the ferment of albuminoid substances is precipitated with the collodion, while the amylaceous ferment remains in the liquor, and may be obtained by evaporation. Calcium phosphate precipitated from an aqueous solution of the pancreatic gland (by phosphoric acid and lime-water), carries down with it the diastasic ferment; while, on the contrary, the albuminoid ferment remains in solution. If, in the experiment made with collodion, the precipitate be redissolved, washed in water, and dried in aqueous ether, we obtain two layers; the lower and watery one contains the albuminoid ferment in solution. As to the ferment which saponifies the fatty matters, it has not yet been obtained by itself, and independently of the two others.

Limits of the Activity of a Soluble Ferment.—Certain

experiments tend to prove that soluble ferments have not an unlimited activity. It is known that, with a determinate weight of one of these bodies, we can modify an incomparably greater quantity of fermentable matter; but does the ferment always lose its power after a time, and become exhausted? The limit of activity has been determined for only a small number of these ferments, and the results published warrant but very undecisive conclusions. Most of the experiments of this kind have been made with impure ferments, still containing albuminoid substances.

Thus, according to Payen and Persoz, one part only of diastase is sufficient to liquefy and turn to sugar 2,000 parts of starch; but, on the one hand, the diastase prepared by alcoholic precipitation, according to the process employed by these observers, is a complex mixture, into which the really active substance enters, perhaps, only in a very small proportion. The ratio of 1 to 2,000 would thus become much smaller. On the other hand, it has been ascertained that the presence of a certain quantity of glucose interferes with the transformation of dextrin; and in order to allow the action to resume its course, it is only necessary to dilute the liquid with much water, or to remove the glucose as it is formed, by causing it to undergo an alcoholic fermentation by means of yeast.

According to M. Berthelot, the alterative ferment of yeast is able to alter from 50 to 100 times its weight of sugar. We ought to make the same remark about these numbers as we did respecting diastase. Nothing, therefore, proves decidedly that soluble ferments lose their specific power as they exercise it. The opposite opinion finds some support in the infinitely small quantities of the ferments necessary to produce a very important and considerable effect. It is also evident that these decompositions are effected rather with disengagement than absorption of heat, since the same decomposition by hydratation takes place by the action of sulphuric acid in small quantities, and the same amount of acid will act indefinitely. They do not, therefore, require the employment and the consumption of living forces, and the principle of the conservation of energy is not opposed to the idea of an indefinitely prolonged action.

This manner of considering the question does not exclude the possible, and even probable fact, shown in certain cases, of progressive weakening of the ferment, after which it will have lost its specific power: a similar chemical decay may accompany the manifestation of the specific power, or be produced without it, in an independent manner, and without being a consequence of it.

Particular Studies of Soluble Ferments or Zymases.— After these general considerations, we will notice what speciality is presented by each indirect ferment. In this particular study, we will not return to the physical characters, the chemical composition and properties of zymases, nor to the method of preparing them in greater or less purity. These different questions have been fully treated in the general study of these bodies, and we should expose ourselves to needless repetitions by bringing forward again for each ferment that which applies to the whole class.

The points which will claim our attention are the chemical reaction, the different sources of the ferment which excite it, and the special conditions which favour it.

1. *Diastases.*—We shall give the general name of diastases to ferments capable of turning starch into sugar.

If we only regard the original and the final terms of the reaction, we arrive at this conclusion, that starch, under the influence of diastase, combines with itself the elements of water, and is converted into glucose.

$$C^6 H^{10} O^5 + H^2 O = C^6 H^{12} O^6$$

Amylaceous matter. Water. Glucose.

Yet, according to the researches of Musculus (Ann. Phys. Chim. (3), vol. 55, p. 203 ; Bulletin de la Soc. Chim., Paris, vol. 23, p. 31, 1874) ; those of Schwarzer (Bull. Soc. Chim., vol. 14, p. 400) ; those of Schultz and Mäerker (Bull. Soc. Chim., vol. 19, p. 17), and those of Payen, the change does not take place so simply. By previously warming the starch with water at 70° C. (158° F.), before adding the diastase, the reaction by means of iodine disappears at the moment that the saccharification has affected a quarter of the liquid.* But the action of diastase does not stop there ; if we add a larger quantity, it continues till it has reached the half (51 per cent., according to Payen ; from 51 to 51·7 per cent., according to Schultz and Mäerker) ; that is to say, that the starch is converted into a mixture of glucose and dextrin in equal equivalents. From this moment the saccharification stops, notwithstanding the addition of greater quantities of diastase.

With soluble starch the action is the same. The

* Not the third, as Musculus had previously announced.

reaction by iodine disappears as soon as a quantity of glucose, equivalent to a quarter of the starch, has been formed; but with an excess of diastase, glucose is rapidly and easily formed at the expense of half the starch.

According to these results, and without taking into consideration the disappearance of the peculiar colour produced by iodine, when the quantity of sugar is equal to one-quarter, a fact which must be interpreted by fresh experiments, we may suppose that the amylaceous molecule is resolved by hydratation into two molecules, one of glucose, the other of dextrin.

$$\underset{\text{Starch.}}{2\,(C^6\,H^{10}\,O^5)} + \underset{\text{Water.}}{H^2\,O} = \underset{\text{Dextrin.}}{C^6\,H^{10}\,O^5} + \underset{\text{Glucose.}}{C^6\,H^{12}\,O^6}$$

M. Payen has shown long since, that by the intervention of yeast we produce the entire transformation of starch into glucose, and finally into alcohol and carbon dioxide.

This experiment would prove that the presence of glucose is antagonistic to the action of diastase upon dextrin.

Thus starch, soluble starch, dextrin, and, according to the researches made by Cl. Bernard, the glycogen matter of animal tissues can be changed into sugar under the influence of diastase.*

* A diastase of considerable activity may be obtained by a modification of the process of MM. Payen and Persoz. One part of germinated barley in powder is added to two parts of water. After an hour's maceration, the liquid is pressed out, and an equal volume of alcohol at 80° C. (176° F.) is added to it. It is then filtered, and the first rather voluminous precipitate is rejected. An equal volume of alcohol is again added to the filtered liquid. A very slight precipitate is then formed, which is collected on a filter; this filter is dried with the precipitate at a gentle heat. A diastase paper, which is very active, is obtained in this manner, and may be indefinitely kept without change. (Musculus.)

Diastasic reactions play an important part in the animal organism, and especially in the digestion of starchy food. Thus we find diastasic ferment in the animal economy as well as in the vegetable tissues. At its first entrance into the digestive tube, the food is mixed with the saliva, a liquid which contains, as shown by M. Mialhe in 1845, *ptyalin*, a fermenting principle analogous, or rather identical, with the diastase of germinated barley. MM. Bouchardat and Sandras proved nearly at the same time the existence of an analogous agent in the pancreatic juice.

The pancreatic juice acts infinitely more energetically than saliva in the saccharification of starchy substances. The action of an infusion of pancreatic juice upon starch is excessively rapid—indeed, we may say, instantaneous; it is, in fact, in the small intestine that the principal digestion of amylaceous substances is effected.

This same ferment is met with in other parts of the organism; everywhere where starch, whether animal or vegetable, must be rendered soluble. Thus, in the liver there exists a kind of animal starch, glycogen, which is turned into sugar by contact with blood, and a ferment, in order to be conveyed in this form into the torrent of the circulation. At the period of animal life when this change is to be accomplished, the ferment appears, and the accumulated starch is destroyed. M. Cl. Bernard (Digestion comparée chez les Animaux et les Végétaux, Revue Scientifique, 1873, p. 515) draws a very striking parallel between the chemical phenomena of nutrition in animals and vegetables.

"The ferment appears in seeds from the commencement of germination; this takes place in the potato in spring; then the fermentiferous agent shows itself in the tubercle, as it did in the germinated barley; it liquefies the starch, and puts it into a condition to be distributed over the parts where it may minister to the nutrition, that is to say, to the development of the life of the plant.

"In most animals, the phase of the production of glycogen, and that of its fermentation, are not so distinc as in plants. The two phenomena are often continuous and simultaneous. An exception must, however, be made with respect to the earliest periods of life, especially in those animals which undergo metamorphoses. For example, if we consider the larva of the common fly, *musca lucelia*, the gentle—to call it by its common name—we find that it contains an enormous quantity of starch. It is truly a bag full of glycogen. At this time we find nothing else but glycogen, and not even a trace of sugar. The reason of this is that the glycose ferment does not yet exist. But soon the chrysalis will succeed to the larva, and then in the new phase of existence, during which the perfect animal is in formation, the reserve of glycogen must be utilized. The ferment appears, and the starch is liquefied.

"Something analogous to this is seen in beings of a higher order, for example, among the mammalia, at that period of the embryonic life when nutrition is hastened, and the plastic and formative activity attains its highest degree. The glycogen matter deposited at various points of the fœtus, and of its envelopes, is set in motion, dissolved, and transformed into sugar. *

* The experiments of Cl. Bernard, Hensen, Magendie, and Schiff prove that in living blood there exists a soluble ferment, capable of transforming starch

"The digestion of starches consists in their transformation into soluble and assimilable substances, soluble in order to be able to circulate from one part of the organism to another. Digestion is, therefore, the prologue of the act of nutrition. Whenever these starchy matters are to nourish an organism, we shall find this previous preparation. But all organisms, in the vegetable as well as in the animal kingdom, employ starchy substances as food, all, therefore, digest these substances, in the strict sense of the word."

The agents in these acts of digestion are the diastase of germinated barley, saliva, pancreatic juice, &c.

The action of diastase on starch is exerted at the ordinary temperature, and attains its maximum effect at 75° C. (167° F.). Boiling destroys the specific power of this substance.

It is almost useless to mention, since the fact is so well known, that the effects of diastase on starch are imitated by the intervention of boiling dilute sulphuric acid.

Alterative Ferment.—Cane, sugar, or saccharose, $C^{12} H^{22} O^{11}$, is transformed, as is proved by the researches of M. Dubrunfant (Ann. de Chim. et de Phys. (3), vol. 21, 169, 1847; Comp. Rend. de l'Acad. vol. 29, p. 51, 1849, and vol. 42, p. 901, 1856) by becoming hydrated, under the influence of dilute mineral and even organic

and glycogen into sugar. This ferment is also found in very fresh liver. It may be eliminated from it by making a cold infusion of a liver, which, on account of general disease, produces no more sugar, and contains no glycogen, and by precipitating it subsequently by alcohol. The deposit, redissolved in water, acts on starch. This special ferment exists in the blood of frogs in spring and in summer. In fact, dextrin injected into their veins passes with the urine in the form of sugar. In winter it is wanting, for dextrin appears in the urine without alteration.

acids, into sugar, which causes the plane of polarization to turn to the left, and energetically reduces Fehling's cupro-potassic reagent. Saccharose, on the contrary, turns the plane of polarization to the right, and has no action on Fehling's liquid. The alteration is produced in the cold, but more rapidly in the warm process; it has been effected by simply boiling solutions of cane-sugar in water, by their exposure to light, or by the mechanical means. Thus the mere pulverizing the sugar causes a small portion of altered sugar to be formed. The product of the alteration is not a unique and simple principle.

The same chemist proved that saccharose, after alteration, is found to be changed into two sugars of the formula $C^6 H^{12} O^6$, one of which turns the plane of polarized light to the right, and is identical with grape-sugar, or ordinary glucose; the other, on the contrary, causes the plane of polarization to deviate to the left; this is the uncrystallizable sugar of acid fruits, or *lævulose.*

As at 15° C. (59° F.), the specific rotatory power of glucose is equal to + 57·8°, whilst that of lævulose is — 106°, and as the two sugars are found in the mixture in equal proportions by weight, it can be understood that this mixture, called altered sugar, has a lævo-rotatory action due to the excess of one of these powers over the other.

Altered sugar has a lævo-rotatory power of 25° at the temperature of 15° C.

The formation of these two sugars (glucose and lævulose) is easily formulated by the following equation:—

$$C^{12}H^{22}O^{11} + H^2O = C^6H^{12}O^6 + C^6H^{12}O^6.$$

Saccharose. Water. Glucose. Lævulose.

This is confirmed by the fact, that under the influence of heat, saccharose is changed into glucose and lævulosan (lævulose less water), which is itself changed into lævulose under the influence of acids.

The following is the manner in which M. Dubrunfant brings about the separation of these two bodies.

Ten grammes of altered sugar are dissolved in 100 grammes of water,* and the whole heated with six grammes of calcium hydrate. The mixture, which at first is fluid, forms into a mass at the end of a short time; this is the calcium lævulosate, which separates as it crystallizes, while the glucosate remains in the mother-liquor. The mass is pressed at once, and the solid and liquid parts, after having been dissolved or diluted with a suitable quantity of water, are decomposed separately by carbon dioxide; the former furnishes a solution of lævulose, the latter of glucose. We can also effect a separation, or at least isolate the lævulose free from glucose, by stopping a fermentation of cane-sugar almost in the middle of its course. The glucose fermenting more easily than the lævulose, the latter is found alone.

Without dwelling any further on all these proofs of a true splitting up, which leave no doubt as to the nature of the reaction, let us proceed to the ferment.

This alteration, this splitting up of the saccharose by

* According to the experiments made at the laboratory of the Sorbonne, the proportion of water indicated by Dubrunfant is too great to allow of the crystallization of the calcium lævulosate. Better results were obtained by employing four parts of water to one part of altered sugar.

hydratation is produced before any alcoholic fermentation takes place. Cane-sugar can not be resolved into alcohol and carbon dioxide till after it has been altered, but beer-yeast itself performs this operation. The action was, at first, attributed to the acidity of the yeast, and M. Pasteur thought that it was the result of the presence of succinic acid.

M. Berthelot was the first to show that the alteration of cane-sugar by yeast is independent of the conditions of acidity and neutrality; that it is due to the intervention of a special soluble ferment contained in the cells of yeast. This principle is found in the water in which yeast has been washed, and acts energetically upon sugar, even in a neutral liquid.

The properties of this soluble ferment, besides its peculiar activity, cause it to resemble all the bodies of this class. I have ascertained that it is easy to separate it entirely from the albuminoid substances which accompany it, by following the method of precipitation by phosphoric acid and lime-water mentioned above.*

Saccharose is not susceptible of being absorbed and assimilated under its original form, any more than amylaceous matter, although it has the advantage over the latter in its solubility. Like starch, it has to be digested, and the products of this digestion are glucose and lævulose.† M. Bernard has proved the existence in the

* M. Béchamp found, after M. Berthelot, the alterative ferment (zymase) in the water of yeast, spontaneously weakened by want of nourishment. Under these conditions, I have myself observed that a liquid is obtained, excessively active, and which produces alteration almost instantaneously.

† Cane-sugar is, according to the expression of M. Bernard, like an inert or indifferent matter which can circulate with impunity in the blood or the sap, without the anatomical elements being able to turn it aside or to appropriate it. He gives as a proof of this, that when cane-sugar is injected into the veins or

intestinal juice of a soluble alterative ferment, similar to that of yeast; it forms one of the most important and useful elements of this secretion. In order to prove this, a solution of pure saccharose (not reducing Fehling's liquid) must be injected into a portion of the intestine included between two ligatures, or placed in contact with an infusion of intestinal mucous membrane; we then see, in a very short time, that the sugar reduces the copper oxide. It may be ascertained by the saccharometer that the deviation of this physiologically altered sugar is the same as that of chemically altered sugar. M. Cl. Bernard has ascertained the existence of alterative ferment in dogs, rabbits, birds, and frogs. M. Balbiani has proved that it is to be found in the digestive tube of silkworms. M. Bernard, by comparing these facts relating to the digestion of saccharose in animals, with the analogous transformations observed in the vegetable economy, as he had before done for the saccharification of starch, shows that the alterative ferment is generally found in all parts of the living economy, and under all the circumstances in which saccharose is necessary for nutrition. Sugar-cane which is flowering, and beet which runs to seed, transform by alteration the sugar laid up in their tissues. The agent is always exactly the same, an alterative ferment. M. Bernard obtained it from beetroot in course of evolution, as M. Berthelot had extracted it from yeast. "Alcoholic fermentation," says our illustrious physiologist, "is a phenomenon correlative to the nutrition of an organism, the yeast of beer, *torula*

cellular tissue of an animal, it is returned, weight for weight, in the urinary excretion, and consequently passes through the economy, without being modified or assimilated.

cerevisiæ. But saccharose is unfit for the nutrition of this microscopic form of life, as it is unsuited to that of higher forms. It is therefore necessary that saccharose should be modified, and transformed into glucose, before it can conduce to the vital changes of the organic ferment. The cell of yeast, by producing this transformation, tends to its own development. It digests the saccharose for itself; alteration is here once more a digestive action of the same nature as those which we have already examined.

"In the same manner as starch, the saccharose, which exists in a state of reserve in the tissues of a great number of plants, is unfit to participate in the nutritive circulation of the plant. It is for this reason that sugar can accumulate and be laid up in store, as occurs in the root of the beet, and in the stem of the sugar-cane. The sugar forms a reserve there, while it waits for the moment to begin to act. This period arrives for beetroot, when it is about to put forth buds, flowers, and fruit; then the sugar diminishes by degrees, and disappears soon after from the tissue and the stalk of the beet, changing into glucose. The leaves at this time contain glucose only; the root yields up its store, and the reserves of sugar which it contained are distributed through the stalk to serve for the purposes of the flower and seed; but this is not possible, except by means of a previous transformation, which changes the chemical nature and the composition of the saccharose and causes it to pass into the state of glucose. This is again a true digestion. Beet, therefore, like yeast, and like animals, must digest its sugar."

I could not resist the pleasure of giving the very words

of our great observer. Nothing can, in fact, give a clearer idea of the importance of the subject of which we are now treating, than the wide scope of the considerations by which these phenomena are connected with the entirety of the great act of nutrition in living organisms. We have not here to deal with isolated chemical reactions, interesting on account of the obscurity which still reigns over the cause which produces them, but with transformations which play an important part in the acts of life and nutrition, the remarkable generality of which is placed in the most striking point of view.

Emulsive and Saponifying Ferment.—M. Cl. Bernard has shown that the pancreatic juice is the only one among all the liquids secreted in the digestive tube, which possesses the remarkable property of forming an emulsion, and afterwards saponifying neutral fatty substances. It constitutes, therefore, the true agent of digestion for fatty substances or natural glycerids.

The emulsion which precedes saponification is rather a physical than a chemical transformation. It is the mechanical division of the fatty liquid which is thus separated into an infinite number of little globules, maintaining themselves by means of a characteristic molecular constitution. Under the microscope, we see a great number of granular bodies swimming in the liquid, and animated by the Brownian movement. Emulsion is the necessary condition for the absorption of fatty bodies.

Among the organic secretions of the digestive canal, the pancreatic juice is the only one which can furnish with oils a complete and persistent emulsion. M.

Bernard attributes this emulsive power to a peculiar soluble ferment. He considers that the persistent emulsion of the fat in milk is due to the presence of an emulsive ferment, and that this nutritive liquid contains the fatty matter ready prepared for absorption.

Emulsion by the pancreatic juice or the extract of the gland is instantaneous. A slower action ensues; this is saponification, that is to say, resolution of trimargarin, triolein, tristearin, &c., by hydratation, into glycerin and fatty acids. This resolution is not always effected in the intestine. The emulsion penetrates into the chyliferous vessels still retaining its lactescent character, and it is not till afterwards, in the different tissues, that saponification takes place. It has been thought that the saponification excited by the pancreatic juice was due to the alkalinity of this liquid. M. Bernard replies to this objection by showing: (1) That other secretions quite as alkaline are inactive; (2) That the tissue of the pancreas, which has no alkaline reaction, rapidly produces phenomena of the same kind; (3) We can destroy by boiling, the specific power of the juice, without weakening its alkalinity. The emulsive ferment may be eliminated, mixed with albuminoid matter, and two other indirect ferments, of which we have already spoken (pancreatic diastase, and the digestive ferment of the albuminoids). The emulsive ferment possesses the property of casein, of being precipitated, in the cold, by magnesium sulphate. It is coagulated by heat.

We find bodies of the same order in plants. If we steep oleaginous seeds in water, we obtain an emulsion in which we soon find glycerin and the free fatty acids. During the germination, the emulsive and saponifying ferment, placed with fat in the presence of water, causes it to undergo true digestion, and renders it assimilable.

The emulsine of almonds has an analogous property; we shall presently see that it acts on certain glucosides. We cannot determine whether this complicated action depends on a mixture of several ferments, or on a variable activity of the same body, according to the elements subjected to it. Kölliker and Müller have shown that the pancreatic juice can effect the decomposition of amygdalin. This is a very important indication.

Perhaps we shall find, when we have succeeded in isolating the emulsive ferment of the pancreatic juice, that it has the same properties as emulsin or synastase.

As a conclusion of that which relates to the fermentation of glycerides, we have only now to obtain definite ideas as to the kind of reactions which take place, by giving the equations of the decompositions, as the result of the labours of M. Berthelot has shown them.

1. Emulsion; a physical change.
2. Splitting up.

$$\underset{\text{Triolein.}}{C^3 H^5 (C^{18} H^{33} O . O)^3} + \underset{\text{Water.}}{3 H^2 O} = \underset{\text{Glycerin.}}{C^3 H^8 O^3} + \underset{\text{Oleic acid.}}{3 (C^{18} H^{33} O . HO)}$$

$$\underset{\text{Tristearine.}}{C^3 H^5 (C^{18} H^{35} O . O)^3} + 3 H^2 O = \underset{\text{Glycerin.}}{C^3 H^8 O^3} + \underset{\text{Stearic acids.}}{3 (C^{18} H^{35} O . HO)}$$

$$\underset{\text{Tributyrin.}}{C^3 H^5 (C^4 H^7 O . O)^3} + 3 H^2 O = \underset{\text{Glycerin.}}{C^3 H^8 O^3} + \underset{\text{Butyric acid.}}{3 (C^4 H^7 O . HO)}$$

Albuminosic Ferment.—Two digestive secretions, the gastric and the pancreatic juices, and probably the intestinal juice also, possess the power of transforming the soluble or insoluble but indiffusible albuminoid substances into soluble and diffusible principles. It has been found that this property is due, as is the case with the digestion of starchy bodies and saccharose, to the intervention of special nitrogenous principles, to which the name of pepsin has been given (the digestive principle of the gastric juice), and to that of the albuminosic ferment of the pancreatic juice.

Pepsin, the method of extracting which we know (processes of Schwann, Wasmann and Vogel, and de Brücke), only acts on albuminoids in an acid medium, and within very restricted limits of temperature. Its action is principally exerted on fibrin. The short time that food continues in the stomach only allows, in any case, very slight transformations; and we must, indeed, admit, that the stomachic, acid, pepsinic digestion of albuminoid substances is, in every respect, a very incomplete operation, and rather preparatory than final.

Notwithstanding all that has been written and said upon the chemical question, but very little is yet known about the real transformations of the albuminoids in the stomach, or under the influence of artificial gastric juice; we will not, therefore, enter into long details on this subject; they would only give us superabundant proofs that everything remains to be done. In fact, until the constitution of albuminoids is definitely ascertained, and we know all the intermediate terms of their transformations and progressive splittings up, we shall not be able to enter on this question with advantage and success. I shall therefore content myself with a few observations.

Action of the Gastric Juice on the Fibrin of Blood.—According to Brücke, uncooked fibrin, placed in contact at 40° C. (104° F.) with a sufficient quantity of gastric juice of good quality, swells and dissolves rapidly, changing at first into a principle soluble in dilute acids, but capable of being precipitated by neutralization ; this principle seems to have all the characters of syntonin, the product of the transformation of muscular tissue under the influence of acids. By prolonged digestion, the syntonin disappears in its turn, and gives place to a unique principle, peptone. It is doubtful whether peptone has time to form in the stomach.

Meissner, on the contrary, asserts that the syntonin, which originates at first, splits up by an ulterior action into an insoluble indigestible substance, parapeptone, not capable of being subsequently transformed by the gastric juice, and into soluble principles (peptones *a b c*). The acid stomachic digestion seems, according to these views, to be a decomposition of proteinic substances. This manner of considering the question appears to me the most probable, and the most conformable to the analogies that we can find, according to what is known of the reactions of the albuminoids.

Other albuminoid substances behave, according to Meissner, nearly in the same manner as fibrin, by becoming previously changed into syntonin.

The digestive action of the pancreatic juice, with reference to albuminoids, is much more energetic and efficacious than that of pepsin. It does not require such limited conditions. Thus the albuminosic ferment of the pancreas develops its power as well in an acid as in an alkaline liquid. The products formed are distinguished from the initial compounds by the property of being no longer precipitated by the neutralization of the liquid ; they are incoagulable and easily diffusible, thus resembling crystalloids. Without wishing to speak very decidedly on this question, I am inclined to consider peptones as bodies closely allied to the diffusable, soluble, uncrystallizable,and incoagulable substances which I have obtained by the incomplete action of hydrated baryta on albumin. As to the characters of the third order, by which an attempt has been made to distinguish the different peptones, we pass them over in silence; they present no definite characters to the mind. We can form a clear idea of the nature of these bodies only by bringing about by hydratation a complete resolution of these bodies into definite chemical principles, such as the elements of urea, tyrosine, leucine, and its homologues, and thus determining their constitution. An investigation of this kind, connected with that which I have undertaken with respect to albuminoid substances, could not fail to produce important results in a physiological point of view.

The analogy of functions led M. Bernard to seek for albuminosic digestion in plants. If, with him, we call by this name every transformation of albuminoid matter into soluble diffusible principles, it is certain that it must exist there.

Thus the phenomena presented by yeast preserved in a damp state without nourishment, and to which we have referred when treating of alcoholic fermentation, ought and must be considered as a true digestion of proteids. It is the same with the continuous chemical transformations which take place in the protoplasm of vegetable and animal cells. Nothing proves that the first products of the change of albuminoids in the organism are entirely excrementitious bodies, and that these products, as long as they remain above a certain limit of splitting up, which they do not attain, as, for example, the crystalloids, such as leucine, tyrosine, &c., can no longer contribute to organic synthesis, and to the formation of new cells or tissues. It is the water with which digested yeast has been washed that contains the most suitable nourishment for the development of this organic ferment; but, excepting the albumin, which is inactive as nourishment for yeast, the nitrogenous principles of the water of yeast are products of an order inferior to that of proteinic substances; they are the results of their splitting up. We do not here speak of leucine and tyrosine, the presence of which has been recognized in the water with which yeast has been washed, and which have no nutritive power, but only of the nitrogenous bodies contained in the uncrystallizable sirup.

Brücke found pepsin in the blood, the muscles, and the urine. This agent, therefore, is not confined to the stomach only; it is diffused over the various parts of the organism, wherever its presence may be necessary, especially where there are albuminoids to be liquefied and digested in order to render them fit for nutrition. Bretonneau had already announced that meat introduced into a sub-cutaneous wound could be digested there as in the stomach.

We have given these details of the indirect fermentations which are connected with the great biological phenomena of digestion, the first act of nutrition in living organisms, because of the great importance that attaches to this subject.

As to the other phenomena of the same order, we shall treat them more slightly, since all that is essential may be described in general terms.

Fermentation of Glucosides.—The ferment which acts most generally and energetically on these bodies is found in sweet and bitter almonds; this is synastase or emulsin. It acts best between 30° and 40° C. (86° and 104° F.); but still it can bear an elevation of temperature to 80° C. (176° F.), without losing its specific power.

Acids and alkalies, in small quantities, do not interfere with the action of emulsin. It is only after having undergone a somewhat advanced state of putrefaction that emulsin loses its qualities. Its intervention may be supplied, up to a certain point, by other principles of animal origin. Substances which are poisonous to plants and to yeast have no influence on the reactions produced by synastase.

All these facts, in addition to its mode of preparation (*see* the process of Wittich, before described), leave no doubt as to the nature of emulsin; it is a soluble ferment of the diastase family.

It will be sufficient to give a short summary of the reactions over which it presides; for further details we must refer to works on chemistry.

1st. Splitting up of amygdalin into glucose, hydrocianic acid, and benzoyl hydride. This reaction explains that which takes place when bitter almonds are pounded with water. The bitter almond contains at the same time amygdalin and emulsin. These two bodies are in distinct cells, and can only react on each other when a mechanical action and solution place them in close contact.

This action may take place within the organism; as, for instance, when we inject dissolved amygdalin and emulsin into the veins of an animal, the subject dies with the symptoms of poisoning by prussic acid.

2nd. Resolution of salicin into glucose and saligenin.
3rd. „ of chlorosalicin into glucose and chlorosaligenin.
4th. „ of helicin into glucose and salicylic hydride.
5th. „ of arbutin into glucose and hydroquinone.
6th. „ of phlorizin into glucose and phloretin.
7th. „ of esculin into glucose and esculetin.
8th. „ of daphnin into glucose and daphnetin.*

Vanillin, the odorous principle of vanilla, has quite recently been artificially produced, by decomposing by

* These decompositions take place according to the following equations:—

$$C^{20}\,H^{27}\,N\,O^{11} + 2\,H^2\,O = 2\,C^6\,H^{12}\,O^6 + C^7\,H^6\,O + C\,N\,H$$
Amygdalin. Water. Glucose. Benzoyl hydride. Prussic acid.

$$C^{13}\,H^{18}\,O^7 + H^2\,O = C^6\,H^{12}\,O^6 + C^7\,H^8\,O^2$$
Salicin. Glucose. Saligenin.

$$C^{13}\,H^{17}\,Cl.\,O^7 + H^2\,O = C^6\,H^{12}\,O^6 + C^7\,H^7\,Cl.\,O^2$$
Chlorosalicin. Glucose. Chlorosaligenin.

$$C^{13}\,H^{16}\,O^7 + H^2\,O = C^6\,H^{12}\,O^6 + C^7\,H^6\,O^2$$
Helicin. Glucose. Salicylic hydride.

$$C^{12}\,H^{16}\,O^7 + H^2\,O = C^6\,H^{12}\,O^6 + C^6\,H^6\,O^2$$
Arbutin. Glucose. Hydroquinone.

$$C^{21}\,H^{24}\,O^{10} + H^2\,O = C^6\,H^{12}\,O^6 + C^{15}\,H^{14}\,O^5$$
Phlorizin. Glucose. Phloretin.

$$C^{21}\,H^{24}\,O^{13} + 3\,H^2\,O = 2\,C^6\,H^{12}\,O^6 + C^9\,H^6\,O^4$$
Esculin. Glucose. Esculetin.

$$C^{31}\,H^{34}\,O^{19} + 2\,H^2\,O = 2\,C^6\,H^{12}\,O^6 + C^{19}\,H^{14}\,O^9$$
Daphnin. Glucose. Daphnetin.

means of synoptase a glucoside contained in coniferous plants (coniferin), and by oxidizing the principle formed by this decomposition.

Emulsin, like most of the soluble ferments, may be replaced, with respect to glucosides, by agents purely chemical. Thus we arrive at the same results, in most cases, by boiling with dilute acids.

A very interesting chemical reaction, of the same order as that which we have observed in bitter almonds, is found in the flour of mustard seed, mixed with water. This mixture, the common mustard plaster, has, as we know, a strong odour and a burning taste. The properties of this sinapism are due especially to the presence of essence of mustard, or allyl sulphocyanide. But this essence does not exist fully formed in the seed any more than the essence of bitter almonds in the kernel; it originates at the expense of a special compound contained in black mustard, under the influence of a soluble ferment, myrosin, contained in white or black mustard, and generally in the seeds of the cruciferæ.

The initial product has been isolated in a crystalline form; this is the potassic salt of a complex acid, myronic acid. As to the ferment, we have nothing of importance to say of it; it may be eliminated from the seeds of the cruciferæ, or of white mustard, which do not contain potassium myronate, by the general methods already described; its physical characters, as well as its composition, are those of all the soluble ferments.

Potassium myronate ($C^{10}\ H^{18}\ N\ .\ K.\ S^{2}\ O^{10}$), whose presence characterizes black mustard, is resolved, under the influence of myrosin, into glucose, allyl sulphocyanide, and potassium, bisulphate, as shown in the following equation :—

$$C^{10}\,H^{18}\,N\,K\,S^{2}\,O^{10} = C^{6}\,H^{12}\,O^{6} + C\,N\,C^{3}\,H^{5}\,I\,S + S\,O^{4}\,K\,H$$

Potassium myronate. Glucose. Allyl sulphocyanide. Potassium bisulphate.

The potassium myronate is extracted in the following manner by the process of M. Bussy, modified by Messrs. Witt and Körner. A kilogramme of black mustard seed pulverized is boiled with 1½ litres of alcohol at 82 per cent., until a quarter of a litre of alcohol is distilled over. It is pressed while hot, and the same operation is repeated with the remainder. This is again pressed, dried at 100° C. (212° F.), pulverized, and digested for twelve hours with three parts of cold water; it is pressed again, and the remainder diffused in two parts of cold water. The watery solutions are evaporated, after the addition of a little barium carbonate, till they are of a sirupy consistence. The remainder is treated with boiling alcohol at 85 per cent. (1 or 1½ litres); it is then filtered, distilled, and left in flat open plates to crystallize.

We will now mention the other analogous fermentations, without dwelling on them.

1. The fresh root of the madder contains alizarin and other colouring matters, insoluble of themselves, under the form of soluble glucosides (Rubian, ruberythric acid). Together with these principles, the same root contains a soluble ferment, erythrozyme, which is not slow in splitting up these glucosides, when the powdered madder-root is mixed with water. Under its influence, the colouring matters, at first dissolved, separate whilst

sugar is formed. M. E. Kopp has succeeded in stopping the action of the ferment, by the addition of a certain quantity of sulphurous acid; he founded on this fact a very interesting process for the extraction of the colouring matters of madder.

2. The infusions of the seeds of buckthorn, when they are kept for some time, also undergo a fermentation which decomposes the soluble colouring glucosides contained in this infusion or decoction; glucose and an insoluble pigment are formed. Boiling with acids produces the same transformation.

3. The biliary acids, the taurocholic and glycocholic, are susceptible, as we know by the works of Strecker, of being resolved by hydratation into taurine and cholalic acid, or into glycocoll and cholalic acid. All that need be done, in order to attain this result, is to boil it for a sufficient time with baryta. The same effect of decomposition is observed when bile is left to itself, to undergo spontaneous change. Taurocholic acid is transformed more easily than its accompanying acid. This easy transformation, due to the intervention of ferments, has for a long time obscured the chemical history of bile.

4. The decomposition of hippuric acid, in the urine of the herbivora, into benzoic acid and glycocoll is a reaction of the same order.

5. This is the case also with two vegetable glucosides, *phillyrin*, contained in the bark of the *Phillyrea latifolia*, and *populin*, from the bark of the aspen. These substances are attacked neither by beer-yeast nor by emulsin; but when they are placed under the conditions of lactic fermentation, they are resolved, the one into glucose and phillygenin—

$$C^{27} H^{34} O^{11} + H^2 O = C^6 H^{12} O^6 + C^{21} H^{24} O^6$$

Phillyrin. Glucose. Phillygenin.

the other into glucose, saligenin, and benzoic acid—

$$C^{13} H^{17} (C^7 H^5 O) O^7 + H^2 O = C^{13} H^{18} O^7 + C^7 H^6 O^2$$

Populin or benzoyl salicin. Salicin. Benzoic acid.

The salicin formed at first is resolved in its turn into glucose and saligenin.

6. Infusions of gall-nuts, left to themselves, ferment; the tannin disappears, and is replaced by gallic and ellagic acid, and glucose.

Strecker, considering the tannin as a glucoside, represented its decomposition by the equation—

$$C^{27} H^{22} O^{17} + 4 H^2 O = C^6 H^{12} O^6 + 3 C^7 H^6 O^5$$

Tannin. Glucose. Gallic acid.

7. According to M. Fremy, pectose is accompanied, in the vegetable tissues in which it is found, by a ferment, pectose, sometimes soluble and at others insoluble, which possesses the property of transforming pectose and pectin into pectic and metapectic acids successively.

Pectic fermentation plays an important part in the conversion of ripe fruits into an over-ripe, half-rotten, or "sleepy" state. It also assists in the formation of vegetable jellies. In fact, the transformation of the natural juices of fruits into jellies is a result of the metamorphosis of pectin, contained in these juices, into the pectosic and pectic acids. If, then, a natural juice—that of the currant, for example—does not contain pectose, we must add to it another juice, or pulp, which contains it, in a soluble or unsoluble form, remembering that boiling, by coagulating the ferment, renders it inactive. Pectic fermentation is effected at about 35° C. (95° F.). (Fremy, Ann. Chim. Phys. (3), vol. 24, p. 1.)

CHAPTER III.

ON THE ORIGIN OF FERMENTS.

THE question of the origin of ferments is intimately connected with that of spontaneous generation. In fact, from the time of Van Helmont and others, who, even in the seventeenth century, gave directions for the production of mice, frogs, eels, &c., the partisans of this mode of generation have, by the progress of the tendency to examine into the causes of things, been driven from the larger animals or plants visible to the naked eye, to the smallest living productions, which we can observe only by the aid of the microscope. But ferments are found among these inferior microscopic organisms. Redi, a member of the Academy of Cimento, showed that the worms in putrefied flesh, which were at first thought to be of spontaneous origin, are only the larvæ from the eggs of flies, and that all that was necessary to prevent entirely the birth of these larvæ, was to surround the decomposing meat with fine gauze; he was the first to ascertain that parasitic animals are sexual and able to lay eggs.

The invention of the microscope, and the numerous observations by which it was followed, towards the end of the seventeenth, and the commencement of the eighteenth, century, gave fresh impulse to the doctrine of spontaneous generation, which had lost all credit in questions concerning the origin of living beings of a higher order.

The question now was how to explain the origin of the various living productions, revealed by the microscope in infusions of vegetable and animal substances, among which no apparent symptom of sexual generation could then be found.

The subject was studied for the first time in a scientific manner by Needham, who published in 1745, in London, a work on this subject. This observer did for infusoria what had already been done for the higher organisms. He protected, or rather endeavoured to protect, vegetable or animal infusions from the action of germs, seeds, or any other agents of mutiplication which could come from without. At the same time he destroyed by a physical agent, heat, the germs which might be supposed to exist beforehand in the liquid. Under these conditions, either living beings will be produced in the midst of the infusion, or none will be found there; in the former case, it must be admitted that these organisms are developed in the medium which is suitable to them, without the intervention of any germ; in the second, that the doctrine of spontaneous generation is false. In reality, the question can only be resolved in this manner, and all experimenters who have entered upon it from Needham's time to the present day ought to have made use of it.

The serious and grave difficulty, on which, during this period, all discussions raised between heterogenists and panspermists have turned, is so to arrange the experiments as to remove every suspicion of the intervention of germs brought from without, or pre-existing in the liquid.

If the result is negative, if when all precautions that seem to be necessary have been taken, and all causes of error have been removed, there is no formation of infusoria, it will be difficult to raise any serious objection to the inevitable conclusion, provided that the methods employed for the purpose of eliminating the pre-existing germs are not of such a nature as to modify the medium, and to render it unfit for the development and the nutrition of living organisms. If, on the contrary, we still meet with the birth of living beings, the suspicion will always revive that the experiment has been badly performed, and that a contrary result would have been obtained by conducting it more carefully. The heterogenists, therefore, find themselves in a more disadvantageous situation than their opponents, and notwithstanding the success which they may obtain, they will never convince them.

We think therefore that it is useless to give here a detailed account of their minute researches; they must be consulted in the original memoirs. *A single experiment which proves, by a negative result, that organic infusions, protected from germs from without, do not give birth to infusoria, is worth more, scientifically speaking, than ten experiments tending to establish the contrary opinion.*

If, therefore, we pass over the details of the fundamental experiments of the heterogenists, and speak of those the results of which are conformable to the ideas of the panspermists, it will not be in a spirit of partiality. We are convinced that the latter are the only ones free from all objections, the relative skill of the operators being disregarded, and considered as nothing in the estimate formed. We may, however, say, that M. Pasteur's researches may serve as a model for all those who may wish to conduct investigations of this kind, whatever may be the preconceived opinion by which they are guided. By their precision, and the care taken to remove every source of error, they leave nothing to be desired.

As the results obtained by M. Pasteur lead him to deny spontaneous generation, his opponents must above all prove that he is mistaken, by adopting the same rigorous experimental conditions.

Needham's experiments, of which we have before spoken, which led this observer to admit and sustain the doctrine of spontaneous generation, consisted essentially in placing organic substances which were capable of decomposition, in vessels hermetically sealed, which were subsequently submitted to a high temperature, in order to destroy the pre-existing germs.

The work of the English writer attracted great notice on account of the support of Buffon, whose ideas he upheld.

Soon after began the great controversy between Needham and Spallanzani, a celebrated Italian physiologist. The latter, in his treatises on animal and vegetable physics, translated by Sennebier, in 1777, refuted by experiment the conclusions arrived at by Needham.

The controversy turned principally on this point; Spallanzani was not satisfied with heating the hermetically sealed vessels containing the infusions, for several minutes, *merely the time which is required to cook a hen's egg, and to destroy the germs,* as Needham expresses it; but he kept them for the space of an hour in boiling water. He then observed no production of infusoria. But, objects the English observer, from the manner in which he treated and put to the torture his nineteen vegetable infusions, it is evident that he not only much weakened, or perhaps totally destroyed, the *vegetative force* of the substances infused, but also entirely corrupted, by the exhalations and the odour of the fire, the small portion of air which remained in the empty part of his vessels. It is not therefore surprising that his infusions, thus treated, gave no signs of life. Such must necessarily have been the case.

This idea, that the action of the temperature of boiling water destroys the vegetative force of infusions, is maintained even at the present day, and has served as an argument to the heterogenists; as they were unable to attack the material correctness of Pasteur's experiments, they did not accept the conclusions which he sought to derive from them.

We find also in the passage just cited, the necessity for the experiments made by Schwann and Helmholtz on calcined air, and for those of Schrœder and F. Dusch, on strained air.

The objection of a possible change in the air contained in the phial, under the influence of prolonged boiling, in presence of organic substances, was a serious one at the time that it was brought forward; it becomes more so, when we know that the air confined over preserved meats, prepared by Appert's process, contains no oxygen. It was, therefore, absolutely necessary to place the infusions in contact with air in a normal condition, after that boiling had deprived them of their pre-existing germs, avoiding at the same time any new germs brought by the air.

For this purpose, Dr. Schwann heated flasks containing the infusions, until the destruction of the germs was insured; but his flask was not closed; it communicated freely with the surrounding air by means of a glass tube bent in the form of an U, and heated, in one part of its length, by means of a bath of fusible alloy. Under these conditions, the air may be renewed in the flasks, but the fresh atmospheric air admitted has undergone, like the infusion, the action of heat, which destroys the germs.

Schwann's experiment was very decisive, as to broth made from meat; and the negative result (no development of infusoria) was quite satisfactory. But it was not the same with analogous trials on alcoholic fermentation, which gave contradictory results.

Ure and Helmholtz repeated and multiplied these experiments with the same success.

To obviate the objection of a possible change by heat, in a mysterious and undefined principle, different from germs, but whose presence in the air was necessary to the production of infusoria, Schultz (Ann. des Sciences Naturelles, vol. 8, (2), 1837) caused the renewed air to pass through energetic chemical re-agents, such as concentrated sulphuric acid. He half filled a glass vessel with distilled water containing various animal and vegetable substances; then stopped the vessel with a cork through which passed two bent tubes, and exposed the apparatus thus arranged to the temperature of boiling water. Then, while the vapour was still escaping through the tubes of which we have just spoken, he adapted to each of them a Liebig's bulb apparatus, one containing concentrated sulphuric acid, and the other concentrated caustic potash. The high temperature must necessarily have destroyed every living thing, all the germs that might happen to be in the inside of the vessel, or of its appendages, and the communication from without was intercepted by the sulphuric acid on one side and the potassa on the other. Nevertheless, it was easy to renew by aspiration at the end of the apparatus which contained the potassa, the air thus enclosed, and the fresh quantities of this fluid which were introduced could not carry with them any living germ, for they were forced to pass through a bath of concentrated sulphuric acid. M. Schultze placed the apparatus thus arranged at a well lighted window, side by side with an open vessel, which contained an infusion of the same organic substances; then he was careful to renew the air in his apparatus several times a day for more than two months, and to examine with the microscope what took place in the infusion. The open vessel was soon found filled with vibrios and monads, to which were soon added polygastric infusoria of a larger size, and even rotifers; but by the most attentive observation he could not discover the least trace of infusoria, confervæ, or mildews, in the infusion contained in the apparatus.

The latest researches of Schrœder and V. Dusch (1854–1859) tended to raise another objection, the possible change in a special principle in the air, by a re-agent as energetic as sulphuric acid. Guided by the experiments of Loëwel, who ascertained that common air, when it had been previously filtered through cotton, was unfit to cause the crystallization of supersaturated solutions of sodium sulphate, they placed one of the tubes of Schutze's apparatus in communication with a tube 3 centimètres (1·18 in.) in diameter, and from 50 to 60 centimètres (19·68 to 23·62 in.) in length, filled with cotton-wool. The other tube was connected with an aspirator.

When the liquid, the interior of the flask, and the tubes, had been been deprived of air by boiling, the apparatus was removed to its place, and the aspiration continued night and day.

The two observers thus proved that meat to which water had been added, the wort of beer, urine, starch, paste, and the various materials of milk taken separately, remained intact in the filtered air.

On the contrary, milk, meat without water, and the yolk of egg, grew putrid as rapidly as in common air.

The result of these experiments is, that there are spontaneous decompositions of organic substances which require nothing but the presence of oxygen gas to cause them to commence; while others need, besides oxygen, the presence in the atmospheric air of *those unknown things*, which are destroyed by heat or sulphuric acid, or are retained by the cotton.

The two observers do not then decide on the nature of these things, and do not assert categorically that they are germs, and, in reality, nothing allows us to draw these conclusions.

M. Pasteur's experiments (Memoire on the Organic Corpuscles which exist in the Atmosphere, Ann. Chim. Phys. (3), vol. 64, p. 27) have advanced the question another step, by proving that they are really germs of ferments and infusoria, which are destroyed by heat, or arrested by the sulphuric acid or cotton in the experiments alluded to above.

M. Pasteur made a hole in a window-shutter, several mètres above the ground, and through this he passed a glass tube half a centemètre ('196 in.) in diameter, and containing a plug of soluble cotton 1 centimètre ('39 in.) in length, kept in its place by a spiral platinum wire. One of the ends of this tube passed into the street; the other communicated with a continuous aspirator. When the air had passed for a sufficient time, the plug of cotton, more or less soiled by the dust which it had intercepted, was placed in a small tube with the mixture of alcohol and ether, which dissolves gun-cotton. It was left for the space of a day. All the dust was deposited at the bottom of the tube, where it is easy to wash it by decantation, without any loss, if care is taken to separate each washing by an interval of repose of from twelve to twenty hours. The deposit, and the liquid which covers it, are put in a watch-glass together; after the evaporation of the alcohol, the remainder is placed in water, and examined with the microscope. M. Pasteur also made use of ordinary sulphuric acid in order to moisten the dust. This agent had the effect of separating the grains of starch and calcium carbonate, which are always found in greater or less quantities in deposits collected on the plug of cotton.

Figs. 24 and 25 represent organic corpuscules, associated wih amorphous particles, as seen through the microscope, under a power of 350 diamètres; the liquid containing them was common sulphuric acid.

Fig. 24 applies to dust collected from the 25th to the 26th of June, 1860; fig. 25 to dust from the very intense fog of January, 1861.

It was not enough to discover with the microscope organic particles mixed with amorphous substances, but

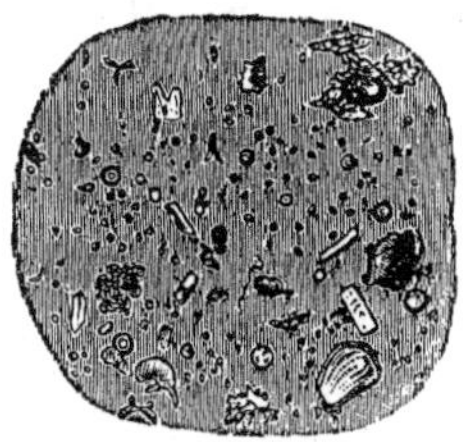

Fig. 24.

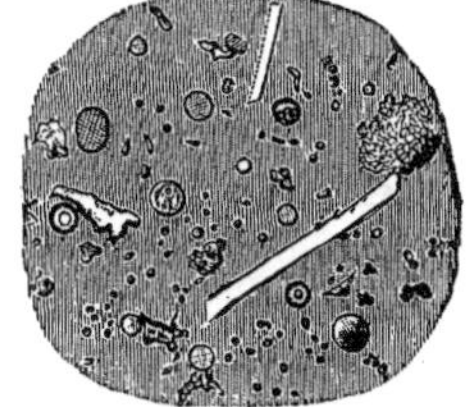

Fig. 25.

Figs. 24 and 25.—Organic corpuscules of dust, mixed with amorphous particles.

it was necessary to prove that these particles really consisted of fertile germs, capable of producing the infusoria which are developed in such abundance in organic liquids exposed to the air.

For this purpose, M. Pasteur arranged the experiment in the following manner:—

Into a flask capable of containing from 250 to 300 cubic centimètres (15 to 18 cub. in.), he introduced 100 or 150 cubic centimètres (6 to 9 cub. in.) of albuminous saccharine water, prepared in the following proportions:—

Water 100
Sugar 10
Albuminoid and mineral matter from beer-yeast ·2 to ·7.

The neck of the drawn-out neck flask communicated with a platinum tube, as shown in Fig. 26. In this first stage of the experiment the T-shaped tube with three stop-cocks is removed, and its place supplied by a simple india-rubber connecting piece. The platinum tube is raised to a red heat by means of a small gas furnace. The liquid is boiled for two or three minutes, and is then allowed to grow completely cold. It is filled with common air, at the ordinary pressure of the atmosphere, but which has been wholly exposed to a red heat; then the neck of the flask is hermetically sealed.

This, being thus prepared, and detached, is placed in a stove at a constant temperature of about 30° C. (86° F.); it may be kept there for any length of time without the least change in the liquid which it contains. It preserves its limpidity, its smell, and and its weak acid reaction; even a very slight absorption of oxygen is mainly to be observed. M. Pasteur affirms that he never had a single experiment which was arranged as described above, which yielded a doubtful result; while water of yeast mixed with sugar, and boiled for two or three minutes, and then exposed to the air, was already in evident process of decomposition in a day or two, and was found to be filled with bacteria and vibrios, or covered with mucors. These experiments are directly opposed to those of Messrs. Pousset, Mantegazzo, Joly, and Musset.

It is therefore clearly proved that sweetened yeast water, a liquid very liable to be decomposed by the contact of common air, may be preserved for years unaltered when it has been exposed to the action of calcined air, after having been allowed to boil for a few minutes (two or three).*

* M. Pasteur has pointed out a cause of want of success, which has led many experimenters into error; by showing that the mercury of a mercurial

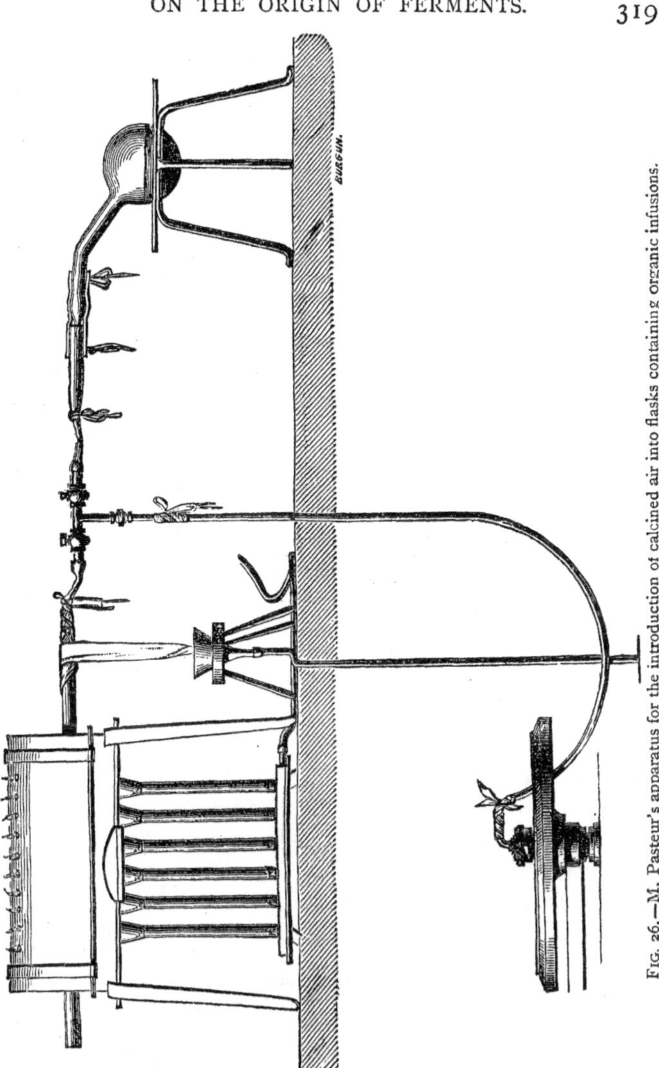

FIG. 26.—M. Pasteur's apparatus for the introduction of calcined air into flasks containing organic infusions.

This being determined, M. Pasteur adapted, by means of an india-rubber tube, the closed point of his flask filled with sweetened yeast water, which had been kept for two or three months in a heated stove, without any development of organisms, to an apparatus arranged like that in figure 26.

The pointed end of the flask passed into a strong glass tube 10 or 12 millimètres (·39 to ·46 in.) in its inner diameter, within which he had placed a piece of tube of small diameter, open at both ends, free to slip in the larger tube, and enclosing a portion of one of the small plugs of cotton loaded with dust. The larger glass tube is bound to a brass tube in form of a T, furnished with stop-cocks, one of which communicates with the air-pump, another with the heated platinum tube, and the third with the flask, by means of the large tube which contains the smaller one with the cotton. These various parts are joined together by means of india-rubber.

The experiment is commenced by exhausting the air, after having closed the stop-cock connected with the red-hot metallic tube. This being afterwards opened, allows calcined air to enter the tubes slowly; this operation (exhaustion and readmission of calcined air) is repeated several times. The point of the flask is then broken off within the india-rubber, and the small tube containing the dust is allowed to slip into the flask, the neck of which is again sealed with the lamp. As an additional proof, and to obviate all objections, the same arrangements were made with similar flasks, prepared

trough is a complete receptacle for living organisms, and consequently that all experiments made with such a trough must necessarily induce a development of infusoria.

like the preceding, but with this difference, that instead of cotton charged with atmospheric dust, there was substituted a small piece of tube containing calcined asbestos; (as an additional precaution, it had been ascertained that calcined asbestos, loaded with atmospheric dust, by the same means as the cotton, gave identical results).

The following are the observations obtained constantly by M. Pasteur:—

In all the flasks, into which dust collected from the air were introduced, 1. Organic productions began to make their appearance in the liquid after 24, 36, or 48 hours at the most. This was precisely the time necessary for the same phenomena to appear in sweetened yeast-water exposed to contact with the atmosphere.

2. The products observed are of the same kind as those which are seen to make their appearance in the liquid when left freely exposed to the air, such as mucors, common mucidines, torulacei, bacteria, and vibrios of the smallest species, the largest of which, the *monas lens*, is only ·004 millimètre (·000157 in.) in diameter.

It was a remarkable circumstance that M. Pasteur never saw any alcoholic fermentation appear, although the composition of the liquid employed was very appropriate to this kind of change.

When the water of yeast is replaced by urine, the experiment being conducted exactly in the same manner, we always notice the absence of any change as long as atmospheric dust has not been introduced, whilst with the addition of this, numerous organisms are developed, in every respect similar to those which appear and are developed in urine kept in the open air.

If, on the contrary, the experiment be repeated with common milk, we may be sure that it will in every case curdle, and become putrid. We shall observe the birth of numerous vibrios of the same species, and bacteria, and the oxygen of the flask will disappear.

M. Pasteur thinks that this result, so different from those observed in other liquids, arises only from the fact that milk contains germs of vibrios which resist the boiling heat of water. To prove this, he boiled milk, not at 100° C. (212° F.), or at the usual pressure of the atmosphere, but at 110° C. (230° F.), under a greater pressure, and he found that the flasks thus prepared, and hermetically sealed, could be kept for an indefinite time in the stove, without giving rise to the smallest production of mould or infusoria. The milk preserves its taste, its smell, and all its properties; and the atmosphere of the flask is only slightly modified in its composition. (After forty days there were still found 18·37 volumes of oxygen per 100 parts of air.)

This difference between milk and urine, or sweetened yeast-water, must be attributed to the alkaline condition of the former medium, whereas the two others are acid.

In fact, if we previously neutralize the acid of the sweetened yeast-water, by means of calcium carbonate, we obtain organisms under the same conditions of the experiment as those under which they were not before developed.

These facts led M. Pasteur to make researches on the comparative action of temperature on the fecundity of

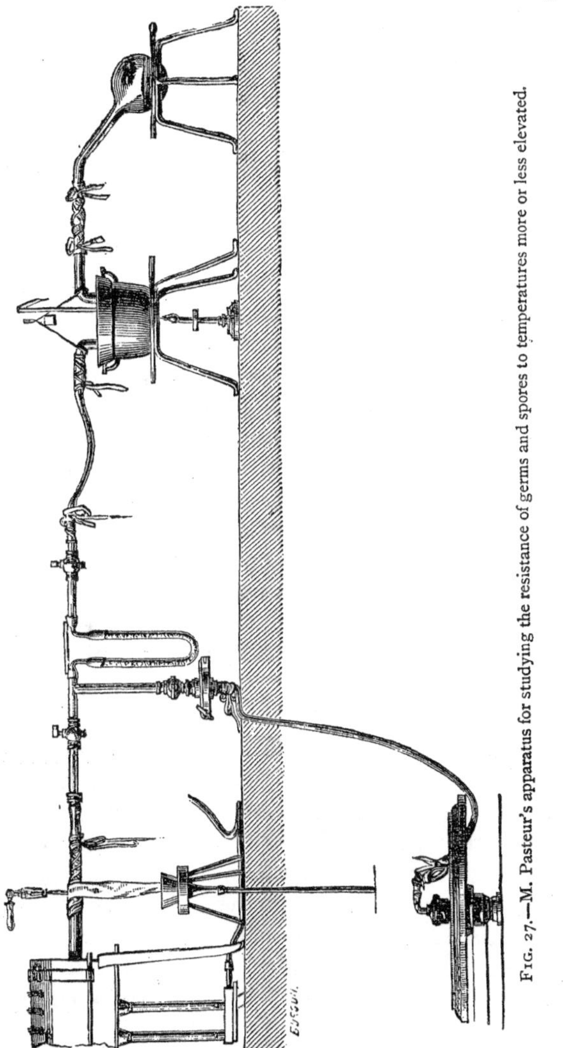

Fig. 27.—M. Pasteur's apparatus for studying the resistance of germs and spores to temperatures more or less elevated.

the spores of the mucidines, and of the germs which exist suspended in the atmosphere.

The following is, in few words, the method followed by him. He passed a small portion of asbestos over the small heads of the moulds which he wished to study; he then placed this asbestos, covered with spores, in a small glass tube, which he introduced into an U tube (Fig. 27) of larger diameter, in which the smaller tube could move freely; one of the extremities of the U tube is joined by india-rubber to a metal tube in form of a T, with stop-cocks. One of these cocks communicated with the air-pump, another with a red-hot platinum tube. The other extremity has an india-rubber tube which is connected with the flask into which the spores are to be introduced; this flask is hermetically sealed, and has been filled with calcined air, and suitable nutritious liquid previously raised to the boiling point. Finally, the U tube dips into a bath of oil, of common water, or salt water, according to the temperature which we wish to attain. Between the U tube and that of platinum, there is a drying tube with sulphuric pumice-stone. When all the apparatus which precedes the platinum tube has been filled with calcined air, and the spores have been maintained at the desired temperature for a sufficient time, which may be varied at pleasure, the point of the flask is broken with a blow of a hammer, without unfastening the india-rubber connecting pieces which attach the flask to the U tube; then inclining to a proper angle, this latter tube, when removed frum its bath, the asbestos with its spores is slipped into the flask. The flask is then hermetically sealed, and is carried to the stove at 20° or 30° C. (68° to 86° F.). The experiment with the dust from the air is also made in the same manner with asbestos.

Without any humidity, the fecundity of the spores of *Penecillium glaucum* is preserved up to 120° C. (248° F.) and even a little above—125° C. (257° F.). It is the same with the spores of the other common mucidines. At 130° C. (266)° F.), the power of developing or multiplying is destroyed in all of them. The limits are the same for the dust from the air.

In all these careful experiments, the most scrupulous precautions were taken to prevent the access of the slightest portion of common air. But, say the partisans of heterogenesis, if the smallest portion of common air develops organisms in any infusion whatever, it must necessarily be the case that, if these organisms are not spontaneously generated, there must be germs of a multitude of various productions in this portion of common air, however small it may be; and if things were so, the ordinary air would be loaded with organic matter which would form a thick mist in it.

M. Pasteur has shown that there is a great deal of exaggeration in the generally received opinion that even the smallest quantity of air is sufficient to develop multitudes of organisms; that, on the contrary, there is not in the atmosphere a continuous cause of these so-called spontaneous generations; that it is always possible to procure, in any determined place, a sufficient but still limited quantity of common air, having undergone no kind of modification, whether physical or chemical, and nevertheless quite unsuited to set up any decomposing action in a liquid eminently putrescible. The method of experimenting is very simple. Into a flask of 250 or 300 cubic centimètres (15 to 18 cub. in.), 150 cubic centimètres (9 cub. in.) of a liquid that has a tendency to decomposition are introduced; the neck of the flask is drawn out with the lamp, leaving the point open; then the liquid is boiled till the vapour escaping from the extremity has expelled all the air; at this moment the point of the flask is closed by the lamp, by means of a blow-pipe, and it is allowed to grow cool. The flask then contains no air; if we break off the point in any particular place, the air re-enters suddenly, carrying into it the germs held in suspension; it is again closed with the lamp, and kept in a stove at a temperature of 20° or 30° C. (68° to 86° F.). In the generality of cases, organisms are developed; these organisms are even more varied than if the liquid were freely exposed to the air, which M. Pasteur explains by saying that, in this case, the germs in small number, in a limited volume of air, are not hindered in their development by germs in greater number or more precocious in their fecundity, which are able to occupy the space, and leave no room for them. But it is especially important to notice in the results obtained by this method, what frequently happens many times in each series of trials, that the liquid continues absolutely intact, however long it may have remained in the stove, as if it had been filled with calcined air. This phenomenon is the more striking, and shows itself in more marked proportions, when the air received into the flasks is taken from a greater height. Thus, among twenty flasks opened in the country, eight contained organic productions; out of twenty opened on the Jura, only five contained any; and out of twenty flasks opened at Montanvert, in a rather high wind, blowing from the deepest gorges of the "Glacier des Bois," one only was affected by any change.

We may also draw other conclusions from this series of observations. Since the putrescible liquid, which had been previously boiled, and which was contained

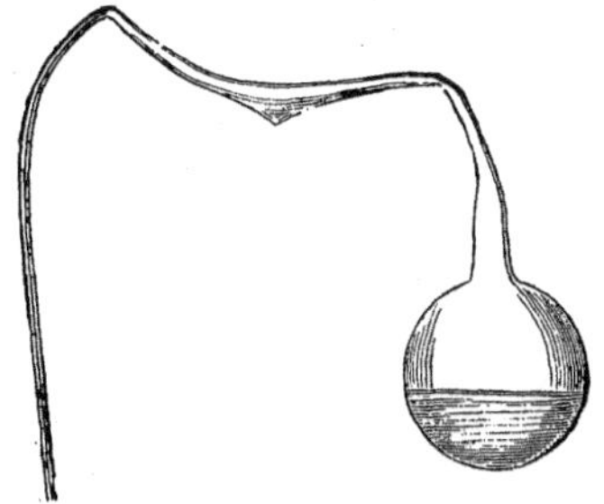

FIG. 28.—M. Pasteur's flask to deprive the air of its germs.

in the flasks, was filled with organic productions in a great number of instances, after the introduction of a limited quantity of air, the genetic power of the infusions had not been destroyed by the material conditions of the experiments. Besides, this objection, which has been raised ever since the earliest controversies between the heterogenists and the panspermists, has been definitely answered by an experiment made by M. Pasteur; he received in a flask, exhausted and deprived of living germs by the momentary application of a sufficiently high temperature, some blood at the instant that it left the organism, and without allowing this liquid, which is so peculiarly putrescible, to come in contact with air. By permitting air deprived of germs, either by calcination or simple filtration, to enter the flask, and then hermetically sealing it, he found that the blood was preserved for an indefinite period intact, although it had not been exposed to heat.

M. Pasteur has also shown that air may be deprived of germs by its passage through a capillary tube bent upon itself. It is, therefore, sufficient, in most cases, to draw out the neck of the flask so as to form a very long narrow tube, which is bent in several directions, as, for example, in Fig. 28. When the air originally contained in it had been expelled, and the pre-existing germs killed by prolonged boiling, the flask is allowed to cool slowly.

In closing our account of M. Pasteur's interesting memoir, in which heterogenesis was driven to its last intrenchments, we must add that this learned chemist endeavoured to deprive his adversaries of one of their principal arguments. Experiments on spontaneous generation have always been conducted with vegetable or animal infusions; it was supposed by Needham, Buffon, and Pouchet that organisms were only thus produced at the moment of expiring nature, when the elements of the beings on which they are developed are entering into new chemical combinations, and are passing fully through the phenomena of fermentation or putrefaction.

In other words, albuminoid matters preserve in some degree a certain reserve of vitality, which would allow them to become organic by contact with oxygen, when the conditions of temperature and humidity are favourable. Starting with the idea that albuminoid substances are only aliments for the germs of infusoria, mucidines, or ferments, M. Pasteur has proved directly that organic substances may be replaced by those which are purely mineral or artificial, or, at least, by substances on which this imaginary vegetative force cannot be supposed to have any influence.

We have spoken elsewhere, at some length, of the experiments made by M. Pasteur and M. Raulin on the nutrition and development of ferments and of mucedines in artificial media composed of pure sugar candy, ammonium tartrate, and phosphates.

The reader will doubtless remember an observation of M. Pasteur's, which we mentioned in passing, but without dwelling upon it. When he introduced into sweetened yeast-water that had been previously boiled and preserved in calcined air atmospheric dust collected at different times and in various places, he never met with alcoholic fermentation. Yet the liquid employed is one of those which are most suitable for the development of alcoholic ferments. The fact may be explained by admitting that the air does not contain spores or germs of *saccharomyces* or apiculated ferment; but then how can we explain the prompt and constant appearance of alcoholic fermentation in the juice of grapes or of fruits in general? The cause of this apparent contradiction is very simple. It is not the air which brings the germs of the alcoholic ferments which propagate and multiply so rapidly in the must of grapes, or, if it brings any, they are in so minute a quantity that they would not be sufficient to set up fermentation in so short a time. These germs are found on the very surface of the fruit, on the grapes which contain the saccharine liquid, the decomposition of which they excite as soon as they are placed in contact with it when the fruit is pressed.

In order to prove this, M. Pasteur prepared a series of forty flasks, with sinuous necks like those which we have described above (Fig. 28), with this difference only, that the neck of the flask, which is drawn out to a fine "swan's neck" is not the only one. Each flask has another straight neck closed by an india-rubber tube furnished with a glass stopper. Into these forty flasks he introduced limpid and filtered grape juice, which, after being boiled, like all liquids which are slightly acid, remains intact, although the extremity of the neck is open. On the other hand, he washed in a few cubic centimètres of water part of a bunch of grapes. Under the microscope, we may perceive in this water the existence of a multitude of organic corpuscules, resembling so closely as to be indistinguishable from them, either spores of minute fungi, alcoholic ferment, or the *Mycoderma vini.* This having been done, M. Pasteur sowed nothing in ten of the forty flasks; in ten others, he placed, by means of the second tube, some drops of the liquid in which the grapes had been washed. In a third series of ten other flasks, he placed some drops of the same liquid, previously raised to the boiling point, and then cooled. Into the ten remaining flasks, he introduced one drop of the juice of a grape taken from those that were not crushed.

The first series gave no product, the juice having remained unaltered. In the second series, there appeared some flakes of mycelium, somc alcoholic ferment, and, afterwards, *Mycoderma vini;* at the end of forty-eight hours the ten flasks were in full fermentation, if they had been experimented upon at the temperature of the air. The third series had not a single flask affected; the liquid remained limpid. In the fourth, a single flask was changed. The conclusion which may be drawn from these facts is simple.* M. Béchamp had already proved, by former experiments, that grapes bear on their surface all that is necessary to cause saccharine water to ferment, even when protected from the air.

With the question of the origin of ferments and of spontaneous generation, is connected another, which may be discussed independently, whatever may be the origin of ferments, whether spontaneous or no; namely, whether a fermenting organism can transform itself into a different ferment, endowed with distinct active properties, when the conditions of its development are modified.

It is evident that this question may be seriously discussed; it is connected with the general development theory which has been applied to higher organisms; it is, therefore, still more applicable to the simplest organisms of the living creation. The facts observed are not entirely opposed to the idea of a transformation of ferments into each other; we have even had occasion to mention some which seem favourable to it. In the meanwhile, these facts are still too few in number to give rise to important inferences; some of them are even contested. We shall therefore merely call attention to this branch of the study of inferior organisms; it has already been the object of valuable researches, which for several reasons deserve to be continued.

* Pasteur, Comp.-Rend., vol. 75, p. 781.

III.

Foods.

By Dr. EDWARD SMITH.

1 vol., 12mo. Cloth. Illustrated. . . . . Price, $1.75.

In making up THE INTERNATIONAL SCIENTIFIC SERIES, Dr. Edward Smith was selected as the ablest man in England to treat the important subject of Foods. His services were secured for the undertaking, and the little treatise he has produced shows that the choice of a writer on this subject was most fortunate, as the book is unquestionably the clearest and best-digested compend of the Science of Foods that has appeared in our language.

"The book contains a series of diagrams, displaying the effects of sleep and meals on pulsation and respiration, and of various kinds of food on respiration, which, as the results of Dr. Smith's own experiments, possess a very high value. We have not far to go in this work for occasions of favorable criticism; they occur throughout, but are perhaps most apparent in those parts of the subject with which Dr. Smith's name is especially linked."—*London Examiner.*

"The union of scientific and popular treatment in the composition of this work will afford an attraction to many readers who would have been indifferent to purely theoretical details. . . . Still his work abounds in information, much of which is of great value, and a part of which could not easily be obtained from other sources. Its interest is decidedly enhanced for students who demand both clearness and exactness of statement, by the profusion of well-executed woodcuts, diagrams, and tables, which accompany the volume. . . . The suggestions of the author on the use of tea and coffee, and of the various forms of alcohol, although perhaps not strictly of a novel character, are highly instructive, and form an interesting portion of the volume."—*N. Y. Tribune.*

IV.

Body and Mind.

THE THEORIES OF THEIR RELATION.

By ALEXANDER BAIN, LL. D.

1 vol., 12mo. Cloth. . . . . Price, $1.50.

PROFESSOR BAIN is the author of two well-known standard works upon the Science of Mind—"The Senses and the Intellect," and "The Emotions and the Will." He is one of the highest living authorities in the school which holds that there can be no sound or valid psychology unless the mind and the body are studied, as they exist, together.

"It contains a forcible statement of the connection between mind and body, studying their subtile interworkings by the light of the most recent physiological investigations. The summary in Chapter V., of the investigations of Dr. Lionel Beale of the embodiment of the intellectual functions in the cerebral system, will be found the freshest and most interesting part of his book. Prof. Bain's own theory of the connection between the mental and the bodily part in man is stated by himself to be as follows: There is 'one substance, with two sets of properties, two sides, the physical and the mental—a *double-faced unity.*' While, in the strongest manner, asserting the union of mind with brain, he yet denies 'the association of union *in place*,' but asserts the union of close succession in time,' holding that 'the same being is, by alternate fits, under extended and under unextended consciousness."'—*Christian Register.*

V.

The Study of Sociology.

By HERBERT SPENCER.

1 vol., 12mo. Cloth. . . . . Price, $1.50.

"The philosopher whose distinguished name gives weight and influence to this volume, has given in its pages some of the finest specimens of reasoning in all its forms and departments. There is a fascination in his array of facts, incidents, and opinions, which draws on the reader to ascertain his conclusions. The coolness and calmness of his treatment of acknowledged difficulties and grave objections to his theories win for him a close attention and sustained effort, on the part of the reader, to comprehend, follow, grasp, and appropriate his principles. This book, independently of its bearing upon sociology, is valuable as lucidly showing what those essential characteristics are which entitle any arrangement and connection of facts and deductions to be called a *science*."—*Episcopalian.*

"This work compels admiration by the evidence which it gives of immense research, study, and observation, and is, withal, written in a popular and very pleasing style. It is a fascinating work, as well as one of deep practical thought."—*Bost. Post.*

"Herbert Spencer is unquestionably the foremost living thinker in the psychological and sociological fields, and this volume is an important contribution to the science of which it treats. . . . It will prove more popular than any of its author's other creations, for it is more plainly addressed to the people and has a more practical and less speculative cast. It will require thought, but it is well worth thinking about."—*Albany Evening Journal.*

VI.

The New Chemistry.

By JOSIAH P. COOKE, Jr.,

Erving Professor of Chemistry and Mineralogy in Harvard University.

1 vol., 12mo. Cloth. . . . . Price, $2.00.

"The book of Prof. Cooke is a model of the modern popular science work. It has just the due proportion of fact, philosophy, and true romance, to make it a fascinating companion, either for the voyage or the study."—*Daily Graphic.*

"This admirable monograph, by the distinguished Erving Professor of Chemistry in Harvard University, is the first American contribution to 'The International Scientific Series,' and a more attractive piece of work in the way of popular exposition upon a difficult subject has not appeared in a long time. It not only well sustains the character of the volumes with which it is associated, but its reproduction in European countries will be an honor to American science."—*New York Tribune.*

"All the chemists in the country will enjoy its perusal, and many will seize upon it as a thing longed for. For, to those advanced students who have kept well abreast of the chemical tide, it offers a calm philosophy. To those others, youngest of the class, who have emerged from the schools since new methods have prevailed, it presents a generalization, drawing to its use all the data, the relations of which the newly-fledged fact-seeker may but dimly perceive without its aid. . . . To the old chemists, Prof. Cooke's treatise is like a message from beyond the mountain. They have heard of changes in the science; the clash of the battle of old and new theories has stirred them from afar. The tidings, too, had come that the old had given way; and little more than this they knew. . . . Prof. Cooke's 'New Chemistry' must do wide service in bringing to close sight the little known and the longed for. . . . As a philosophy it is elementary, but, as a book of science, ordinary readers will find it sufficiently advanced."—*Utica Morning Herald.*

D. APPLETON & CO., Publishers, 549 & 551 Broadway, N. Y.

VII.

The Conservation of Energy.

By BALFOUR STEWART, LL. D., F. R. S.

With an Appendix treating of the Vital and Mental Applications of the Doctrine.

1 vol., 12mo. Cloth. Price, $1.50.

"The author has succeeded in presenting the facts in a clear and satisfactory manner, using simple language and copious illustration in the presentation of facts and principles, confining himself, however, to the physical aspect of the subject. In the Appendix the operation of the principles in the spheres of life and mind is supplied by the essays of Professors Le Conte and Bain."—*Ohio Farmer.*

"Prof. Stewart is one of the best known teachers in Owens College in Manchester.

"The volume of THE INTERNATIONAL SCIENTIFIC SERIES now before us is an excellent illustration of the true method of teaching, and will well compare with Prof. Tyndall's charming little book in the same series on 'Forms of Water," with illustrations enough to make clear, but not to conceal his thoughts, in a style simple and brief."—*Christian Register, Boston.*

"The writer has wonderful ability to compress much information into a few words. It is a rich treat to read such a book as this, when there is so much beauty and force combined with such simplicity.—*Eastern Press.*

VIII.

Animal Locomotion;

Or, WALKING, SWIMMING, AND FLYING.

With a Dissertation on Aëronautics.

By J. BELL PETTIGREW, M. D., F. R. S., F. R. S. E., F. R. C. P. E.

1 vol., 12mo. . . . . . Price, $1.75.

"This work is more than a contribution to the stock of entertaining knowledge, though, if it only pleased, that would be sufficient excuse for its publication. But Dr. Pettigrew has given his time to these investigations with the ultimate purpose of solving the difficult problem of Aëronautics. To this he devotes the last fifty pages of his book. Dr. Pettigrew is confident that man will yet conquer the domain of the air."—*N. Y. Journal of Commerce.*

"Most persons claim to know how to walk, but few could explain the mechanical principles involved in this most ordinary transaction, and will be surprised that the movements of bipeds and quadrupeds, the darting and rushing motion of fish, and the erratic flight of the denizens of the air, are not only anologous, but can be reduced to similar formula. The work is profusely illustrated, and, without reference to the theory it is designed to expound, will be regarded as a valuable addition to natural history." —*Omaha Republic.*

IX.

Responsibility in Mental Disease.

By HENRY MAUDSLEY, M. D.,

Fellow of the Royal College of Physicians; Professor of Medical Jurisprudence in University College, London.

1 vol., 12mo. Cloth. . . Price, $1.50.

"Having lectured in a medical college on Mental Disease, this book has been a feast to us. It handles a great subject in a masterly manner, and, in our judgment, the positions taken by the author are correct and well sustained."—*Pastor and People.*

"The author is at home in his subject, and presents his views in an almost singularly clear and satisfactory manner. . . . The volume is a valuable contribution to one of the most difficult, and at the same time one of the most important subjects of investigation at the present day."—*N. Y. Observer.*

"It is a work profound and searching, and abounds in wisdom."—*Pittsburg Commercial.*

"Handles the important topic with masterly power, and its suggestions are practical and of great value."—*Providence Press.*

X.

The Science of Law.

By SHELDON AMOS, M.A.,

Professor of Jurisprudence in University College, London; author of "A Systematic View of the Science of Jurisprudence," "An English Code, its Difficulties and the Modes of overcoming them," etc., etc.

1 vol., 12mo. Cloth. . . . . Price, $1.75.

"The valuable series of 'International Scientific' works, prepared by eminent specialists, with the intention of popularizing information in their several branches of knowledge, has received a good accession in this compact and thoughtful volume. It is a difficult task to give the outlines of a complete theory of law in a portable volume, which he who runs may read, and probably Professor Amos himself would be the last to claim that he has perfectly succeeded in doing this. But he has certainly done much to clear the science of law from the technical obscurities which darken it to minds which have had no legal training, and to make clear to his 'lay' readers in how true and high a sense it can assert its right to be considered a science, and not a mere practice."—*The Christian Register.*

"The works of Bentham and Austin are abstruse and philosophical, and Maine's require hard study and a certain amount of special training. The writers also pursue different lines of investigation, and can only be regarded as comprehensive in the departments they confined themselves to. It was left to Amos to gather up the result and present the science in its fullness. The unquestionable merits of this, his last book, are, that it contains a complete treatment of a subject which has hitherto been handled by specialists, and it opens up that subject to every inquiring mind. . . . To do justice to 'The Science of Law' would require a longer review than we have space for. We have read no more interesting and instructive book for some time. Its themes concern every one who renders obedience to laws, and who would have those laws the best possible. The tide of legal reform which set in fifty years ago has to sweep yet higher if the flaws in our jurisprudence are to be removed. The process of change cannot be better guided than by a well-informed public mind, and Prof. Amos has done great service in materially helping to promote this end."—*Buffalo Courier.*

XI.

Animal Mechanism,

A Treatise on Terrestrial and Aërial Locomotion.

By E. J. MAREY,

Professor at the College of France, and Member of the Academy of Medicine.

With 117 Illustrations, drawn and engraved under the direction of the author.

1 vol., 12mo. Cloth. . . . . Price, $1.75

"We hope that, in the short glance which we have taken of some of the most important points discussed in the work before us, we have succeeded in interesting our readers sufficiently in its contents to make them curious to learn more of its subject-matter. We cordially recommend it to their attention.

"The author of the present work, it is well known, stands at the head of those physiologists who have investigated the mechanism of animal dynamics—indeed, we may almost say that he has made the subject his own. By the originality of his conceptions, the ingenuity of his constructions, the skill of his analysis, and the perseverance of his investigations, he has surpassed all others in the power of unveiling the complex and intricate movements of animated beings."—*Popular Science Monthly.*

XII.

History of the Conflict between Religion and Science.

By JOHN WILLIAM DRAPER, M. D., LL. D.,

Author of "The Intellectual Development of Europe."

1 vol., 12mo. . . . . Price, $1.75.

"This little 'History' would have been a valuable contribution to literature at any time, and is, in fact, an admirable text-book upon a subject that is at present engrossing the attention of a large number of the most serious-minded people, and it is no small compliment to the sagacity of its distinguished author that he has so well gauged the requirements of the times, and so adequately met them by the preparation of this volume. It remains to be added that, while the writer has flinched from no responsibility in his statements, and has written with entire fidelity to the demands of truth and justice, there is not a word in his book that can give offense to candid and fair-minded readers."—*N. Y. Evening Post.*

"The key-note to this volume is found in the antagonism between the progressive tendencies of the human mind and the pretensions of ecclesiastical authority, as developed in the history of modern science. No previous writer has treated the subject from this point of view, and the present monograph will be found to possess no less originality of conception than vigor of reasoning and wealth of erudition. . . . The method of Dr. Draper, in his treatment of the various questions that come up for discussion, is marked by singular impartiality as well as consummate ability. Throughout his work he maintains the position of an historian, not of an advocate. His tone is tranquil and serene, as becomes the search after truth, with no trace of the impassioned ardor of controversy. He endeavors so far to identify himself with the contending parties as to gain a clear comprehension of their motives, but, at the same time, he submits their actions to the tests of a cool and impartial examination."—*N. Y. Tribune.*

XIII.

THE DOCTRINE OF

Descent, and Darwinism.

By OSCAR SCHMIDT,

Professor in the University of Strasburg.

WITH 26 WOODCUTS.

1 vol., 12mo. Cloth. . . . . Price, $1.50.

"The entire subject is discussed with a freshness, as well as an elaboration of detail, that renders his work interesting in a more than usual degree. The facts upon which the Darwinian theory is based are presented in an effective manner, conclusions are ably defended, and the question is treated in more compact and available style than in any other work on the same topic that has yet appeared. It is a valuable addition to the 'International Scientific Series.'"—*Boston Post.*

"The present volume is the thirteenth of the 'International Scientific Series,' and is one of the most interesting of all of them. The subject-matter is handled with a great deal of skill and earnestness, and the courage of the author in avowing his opinions is much to his credit. . . . This volume certainly merits a careful perusal."—*Hartford Evening Post.*

"The volume which Prof. Schmidt has devoted to this theme is a valuable contribution to the Darwinian literature. Philosophical in method, and eminently candid, it shows not only the ground which Darwin had in his researches made, and conclusions reached before him to plant his theory upon, but shows, also, what that theory really is, a point upon which many good people who talk very earnestly about the matter are very imperfectly informed."—*Detroit Free Press.*

XIV.

The Chemistry of Light and Photography;

In its Application to Art, Science, and Industry.

By Dr. HERMANN VOGEL,

Professor in the Royal Industrial Academy of Berlin.

WITH 100 ILLUSTRATIONS.

12mo. . . . . Price, $2.00.

"Out of Photography has sprung a new science—the Chemistry of Light—and, in giving a popular view to the one, Dr. Vogel has presented an analysis of the principles and processes of the other. His treatise is as entertaining as it is instructive, pleasantly combining a history of the progress and practice of photography—from the first rough experiments of Wedgwood and Davy with sensitized paper, in 1802, down to the latest improvements of the art—with technical illustrations of the scientific theories on which the art is based. It is the first attempt in any manual of photography to set forth adequately the just claims of the invention, both from an artistic and a scientific point of view, and it must be conceded that the effort has been ably conducted."—*Chicago Tribune.*

XV.

Fungi;

THEIR NATURE, INFLUENCE, AND USES.

By M. C. COOKE, M. A., LL. D.

Edited by Rev. M. J. BERKELEY, M. A., F. L. S.

With 109 Illustrations. Price, $1.50.

"Even if the name of the author of this work were not deservedly eminent, that of the editor, who has long stood at the head of the British fungologists, would be a sufficient voucher for the accuracy of one of the best botanical monographs ever issued from the press. . . . The structure, germination, and growth of all these widely-diffused organisms, their habitats and influences for good and evil, are systematically described."—*New York World.*

"Dr. Cooke's book contains an admirable *résumé* of what is known on the structure, growth, and reproduction of fungi, together with ample bibliographical references to original sources of information."—*London Athenæum.*

"The production of a work like the one now under review represents a large amount of laborious, difficult, and critical work, and one in which a serious slip or fatal error would be one of the easiest matters possible, but, as far as we are able to judge, the new hand-book seems in every way well suited to the requirements of all beginners in the difficult and involved study of fungology."—*The Gardener's Chronicle* (*London*).

XVI.

The Life and Growth of Language:

AN OUTLINE OF LINGUISTIC SCIENCE.

By WILLIAM DWIGHT WHITNEY,

Professor of Sanskrit and Comparative Philology in Yale College.

1 vol., 12mo. Cloth. Price, $1.50.

"Prof. Whitney is to be commended for giving to the public the results of his ripe scholarship and unusually profound researches in simple language. He draws illustrations and examples of the principles which he wishes to impart, from common life and the words in frequent use.

"The topics discussed in this volume are, for the most part, those which have been already treated by other writers on philology, and even by the author himself, in his volume on 'Language, and the Study of Language,' published a few years ago, and, though many of the truths here set forth are those with which students in the same line of investigation are generally familiar, all will rejoice to see them restated in such a fresh and simple way.

"This work, while valuable to scholars, will be interesting to every one."—*The Churchman.*

"This work is an important contribution to a science which has advanced steadily under conditions that appear constantly to throw an increasing light on difficult questions, and at each step clear the way for further discoveries."—*Chicago Inter-Ocean.*

"Prof. Whitney is undoubtedly one of the foremost of English-speaking philologists, and occupies an enviable position in the wider circle of European students of language.

"His style, clear, simple, picturesque, abounding in striking illustrations, and apt in comparisons, is admirably fitted to be the vehicle of a popular treatise like the work under consideration."—*Portland Daily Press.*

D. APPLETON & CO., PUBLISHERS, 549 & 551 Broadway, N. Y.

www.ingramcontent.com/pod-product-compliance
Lightning Source LLC
LaVergne TN
LVHW021252110826
845151LV00002BA/592
9781425535445